T0133602

Solar PV Engin
Installation

'Sean's book was instrumental to my passing the NABCEP exam. It really hit the nail on the head.'

– Robert Wylie Hyde, Kaplan Clean Tech PV Instructor, NABCEP Certified PV Installation Professional

'I comfortably passed the NABCEP Exam and I couldn't have done it without Sean's book!'

– Tim Oyler, NABCEP Certified PV Installation Professional

The NABCEP PV Installation Professional (PVIP) Exam is the gold standard certification exam for PV professionals and is famously difficult to pass. As the industry grows and jobs have become more specialized, the Board has responded to this with NABCEP Specialist Certification Exams.

As well as the content of the PVIP, this book also covers the material in the Design, Installer and Commissioning & Maintenance Specialist Certification Exams, and provides test-taking strategy that can be used to most effectively study for and pass these assessments. Code and theory is explained in the first half of this guide, acting as a useful background for the second half, which consists of practice exam questions and answers, complete with detailed explanations. It also contains essential strategy tools, short-term memory tips and recommended reading which will be invaluable for anyone studying for the exams. The material covered in this book is not limited to those taking the test, but will also act as a valuable tool for career progression, helping the reader to work safely with code-compliant PV systems.

This second edition has been fully revised and updated to reflect the new developments in solar and energy storage systems and new technologies.

Sean White is an experienced teacher, instructor and professional in solar PV based in the USA. He received the 2014 IREC 3i Award (Innovation. Ingenuity. Inspiration) in the category of Clean Energy Trainer of the Year and the 2011 Mike Holt Enterprises 1st place Instructor award, and was on the NABCEP PV Installation Professional Technical Committee. He is the author of *Solar Photovoltaic Basics, 2nd edition* (Routledge, 2018), *Photovoltaic Systems and the National Electrical Code* (Routledge, 2018, with Bill Brooks) and *PV Technical Sales* (Routledge, 2016).

Solar PV Engineering and Installation

Preparation for the NABCEP PV Installation Professional, Specialist and Inspector Certification Exams

Sean White

Second edition published 2019
by Routledge
2 Park Square, Milton Park, Abingdon, Oxon OX14 4RN

and by Routledge
52 Vanderbilt Avenue, New York, NY 10017

Routledge is an imprint of the Taylor & Francis Group, an informa business

© 2019 Sean White

The right of Sean White to be identified as author of this work has been asserted by him in accordance with sections 77 and 78 of the Copyright, Designs and Patents Act 1988.

All rights reserved. No part of this book may be reprinted or reproduced or utilised in any form or by any electronic, mechanical, or other means, now known or hereafter invented, including photocopying and recording, or in any information storage or retrieval system, without permission in writing from the publishers.

Trademark notice: Product or corporate names may be trademarks or registered trademarks, and are used only for identification and explanation without intent to infringe.

First edition published by Routledge 2015

British Library Cataloguing-in-Publication Data
A catalogue record for this book is available from the British Library

Library of Congress Cataloging-in-Publication Data
Names: White, Sean (Electrical engineer), author.
Title: Solar PV engineering and installation : preparation for the NABCEP PV installation professional, specialist and inspector certification exams / Sean White.
Description: Second edition. | New York, NY : Routledge, [2019] | Includes index.
Identifiers: LCCN 2019008203 (print) | LCCN 2019010775 (ebook) | ISBN 9780429435553 (Master) | ISBN 9781138348578 (hardback) | ISBN 9781138348592 (pbk.) | ISBN 9780429435553 (ebook)
Subjects: LCSH: Photovoltaic power systems—Examinations—Study guides.
Classification: LCC TK1087 (ebook) | LCC TK1087 .W46 2019 (print) | DDC 621.31/244—dc23
LC record available at https://lccn.loc.gov/2019008203

ISBN: 978-1-138-34857-8 (hbk)
ISBN: 978-1-138-34859-2 (pbk)
ISBN: 978-0-429-43555-3 (ebk)

Typeset in Rotis Sans Serif
by Apex CoVantage, LLC

Contents

Preface

This book is written for those who would like an advanced knowledge of solar PV systems and to prepare people for the NABCEP PV Installation Professional (PVIP) exam, which is the gold standard certification exam for PV professionals. This book is also designed to prepare students to sit for the NABCEP PV Specialist exams, including the NABCEP PV Design Specialist (PVDS) exam, the NABCEP PV Installation Specialist (PVIS) exam, and the NABCEP PV Commissioning & Maintenance Specialist (PVCMS) exams. This book will also help people study for the NABCEP PV Inspector exam and other regional contractors licensing exams. Perhaps the difference between a test question on a PVDS, PVIS, or PVCMS exam is the wording and more so than the knowledge required to correctly answer the question.

The PVIP exam is by far the most popular of all of these exams. It is also highly recommended that anyone would start their studies by taking the NABCEP Associate Exam, which is covered in my book *Solar Photovoltaic Basics*. Even if you do not intend to take an exam, the material that you learn from this book will be invaluable, help you further your career, salary, help you efficiently work safely with code-compliant PV systems, and possibly even save lives.

The pass rate of the NABCEP PV certification exams is known to be low and the exams are famously difficult, which is a good thing for those who pass, meaning that they are known in the industry to understand PV systems inside and out and are in high demand around the world. This book is designed to give direction to study the material that can be learned. There is also material on the exams, which is difficult to learn due to the vast amount of material covered. This book will provide test-taking strategy that can be used to most effectively study for and pass the exam.

Everyone in the PV industry knows someone who has failed the NABCEP PV Installation Professional Exam at least once and many of us know intelligent people who have failed the exam multiple times. Using this book, along with long hours of study, is the best way to ensure an 8.5" × 11" NABCEP PV Installation Professional Certificate shows up in your mailbox.

The heart of this book is the detailed explanations of practice exam questions. We explain code and theory in the first half of the book, but putting everything we have learned to practical use is the best way of learning, retaining, and knowing how to perform calculations necessary for passing the exam and being a successful NABCEP Certified PV Professional.

Before reading this book, you should either have a good basic knowledge of the basics of PV from reading and studying *Solar Photovoltaic Basics: A Study Guide for the NABCEP Associate Exam* or from equivalent studies. This book covers the NEC, but for more in-depth NEC studies, read *Photovoltaic Systems and the National Electrical Code* by Sean White and Bill Brooks. These books were written to be easy to read and to fit easily in your bag or big pocket without breaking your pocketbook. Studying this material is the best way to efficiently pass exams.

No matter how good someone is at reading plans, following directions, and wiring code-compliant safe and efficient PV systems, they still need to understand *how to use the NEC Code Book* in order to pass these exams.

You can read this book in one day. But there is a difference between reading and studying. The best way to use this book is to study it in detail and to go over it many times until you can retain the information and concepts in it. You should also spend time studying with this book and the NEC at the same time, so you can learn to use the NEC for PV systems.

Different strategies for studying:

1. Study for years in order to pass the exam and read every book you can. Years ago, four years of experience installing PV systems was required in order to sit for the exam. This method is the best method for passing the exam.
2. Study as much as you can and if you are experienced at studying and perhaps an electrician or an engineer and you read this book, you will have a chance at passing the exam on your first try.
3. You are very busy installing and designing PV systems and you signed up to take the exam, but have not had as much time to study as you thought you would. You may not pass the exam on your first try, but it is a good

idea to take the exam to get a feel for it, which means you will have a better chance at passing it next time around. This is a good book to cram with and improve your score.

No book can contain the vast amount of information that can be covered on the exam. However, I believe that studying with this book will reap the maximum amount of correct answers per hour of study time.

Sean White

Certain tables in this work are reprinted with permission from NFPA 70®-2017, *National Electrical Code®*, Copyright © 2016, National Fire Protection Association, Quincy, MA. This reprinted material is not the complete and official position of the NFPA on the referenced subject, which is represented only by the standard in its entirety.

Memorize and familiarize

This brief chapter is loaded with material that will help you prepare for the exam and has some of the most important material that you should be able to access in the NEC and (hopefully) your mind. It is recommended to study this chapter on a regular basis and to commit as much as you can to memory.

> There is so much information in the NEC, it is sometimes more difficult to know which articles *not* to use, rather than to know what *to* use. This book will steer you to the common articles that are used in designing PV systems.

IMPORTANT NEC PV ARTICLES AND SECTIONS

690	PV systems
705	Interconnections (includes ac microgrids)
706	Energy storage systems (ESS)
710	Standalone systems
712	Direct current microgrids
Chapter 1	General
100	Definitions
110	Requirements for general installations
110.14(C)	Terminal temperatures
110.16	Arc-flash hazard warning
110.21(B)	Field applied hazard markings (ANSI references)
110.26(A)(1)	Working spaces

110.28	Enclosure selection (NEMA enclosures)
Chapter 2	Wiring and protection
200	Use and identification of grounded conductors
200.6	6AWG and smaller can be marked white for PV
230	Services
240	Overcurrent protection
240.4(D)	Small conductor rule
240.6	Standard fuse and breaker sizes
250	Grounding and bonding
250.52	Electrodes
250.52(A)	Electrodes permitted
250.53	Grounding electrode system installation
250.66	AC GEC
250.166	DC GEC
250.122 (Table)	ECG (ac and dc)
Chapter 3	Wiring methods and materials
300.5 (Table)	Minimum cover requirements (how deep to bury conduit)
300.7	Raceways exposed to different temperatures
310	Conductors for general wiring (Wire sizing)
310.15(B)(2)(a)	Ambient temperature correction factors
310.15(B)(3)(a)	Adjustment for >3 current carrying conductors in conduit
310.16	Conductor ampacity in raceway cable or buried
310.17	Conductor ampacity in free air
314	Junction boxes, enclosures, outlets
320–362	Raceways (conduit) and cables
330	Metal-clad: MC cable
334	Non-metallic sheathed cable: NM, NMC and NMS
338	Service entrance cable: SE, USE (USE-2)
342	Intermediate metal conduit: IMC
344	Rigid metal conduit: RMC
352	Rigid PVC
352.44	Expansion joints required for rigid PVC if > ¼" expansion
356	Liquid tight flexible non-metallic conduit: LFNC
358	Electrical metallic tubing: EMT
Chapter 4	Equipment for general use
480	Batteries (also in articles 706, 710, and 712)
Chapter 6	Special equipment
690.2	Definitions

690.7 Maximum voltage
690.8 Circuit sizing and current
690.9 Overcurrent protection
690.11 Arc-fault protection (dc)
690.12 Rapid shutdown
690.13–.15 Part III Disconnecting means
690.31–.34 Part IV Wiring methods
690.41–.50 Part V Grounding
690.51–.56 Part VI Marking
Article 691 Large-scale photovoltaic (PV) electric power production facility
Chapter 7 Special conditions
705.10 Directory
705.12 Point of connection
705.12(A) Supply-side
705.12(B) Load side
705.12(B)(2)(1) Feeders
705.12(B)(2)(2) Taps
705.12(B)(2)(3) Busbars
705.12(B)(2)(3)(a) 100% rule
705.12(B)(2)(3)(b) 120% rule
705.12(B)(2)(3)(c) Sum rule (loads + breakers ≤ busbar)
705.31 Location of supply-side connection disconnect ≤ 10 ft
705.32 Connect to supply-side of ground-fault protection (with exception)

CHAPTER 9: TABLES

Tables 1, 4, and 5 used for determining how many wires fit into conduit:

Table 1 Percentage cross-section of conduit
Table 4 Internal area of conduit for wires
Table 5 Dimensions of conductors

Table 8 for determining voltage drop

Table 8 Conductor properties (used for voltage drop calculations)

FLASH CARD MATERIAL

- Area of circle is **3.14 × radius^2**
- Area of cylinder is **depth × 3.14 × radius^2**
- Length × width × height = volume
- 2.54 cm/inch
- 3.28 feet per meter
- Wind category B is urban/suburban or wooded with close obstructions (typical)
- Wind category C is scattered obstructions
- Wind category D is no obstructions and wide open
- Wind category A is no longer in use
- Roof zone 3 = corners
- Roof zone 2 = edges
- Roof zone 1 = middle (best for solar)
- PV source circuit often called a string
- Dc-to-dc converter often called an optimizer
- PV module often called a panel (technically incorrect)
- PV source circuit is between PV and combiner
- PV output circuit is after parallel connections in combiner
- Dc-to-dc converter source circuits are between dc-to-dc converters
- PV maximum current definition = Isc × 1.25
- PV required ampacity for continuous current = Isc × 1.56
- Required ampacity for continuous current for other than PV source and output circuits is 125% of current
- 156% ampacity correction only used for PV source and output circuits
- 156% not used with conditions of use; use 125% and compare
- Do not calculate continuous current and conditions of use in same calculation
- OCPD = overcurrent protection device = fuse or circuit breaker
- OCPD size is maximum circuit current × 1.25 and round up (under 800A)
- If disconnects are in different locations, then directory or plaque required at each disconnect
- PV wire used for ungrounded arrays in free air
- Round-down string sizing for cold temperature high Voc calculations
- Round-up string sizing for hot temperature low Vmp calculations
- Rapid shutdown array boundary 1 foot outside and 3 feet inside building
- Pilot hole 67–80% of lag bolt size

- 3 ft space at roof ridge for fire department usually
- For fine-stranded cable use fine-stranded compatible lugs
- No different orientations within PV source circuits
- Different orientations acceptable in dc-to-dc converter source circuits
- Width of working space = width of equipment or 30" whichever is greater
- Time of use rates are usually more expensive on summer afternoons
- Tiered utility rates are more expensive as you use more
- Sine wave is less harmonic distortion
- Conductors in conduit outside have to be wet-rated
- Array gets 80% STC irradiance (current) at 800 watts per square meter
- ASHRAE can give temperatures and wind speeds for an area
- Multimodal inverter, utility shuts down, power critical loads
- A grounded conductor on an insulated lug is good for measuring voltage
- Ground fault protection required for PV on buildings
- Insulation testing good for checking intermittent ground faults
- Insulation testing is done with a megohmmeter (often called megger)
- A grounded conductor is always white or gray in North America (blue in much of world)
- Do not have to upsize EGC when upsizing current carrying conductors for voltage drop
- Ground rods are often copper coated steel 5/8" diameter 8 feet long
- Do not use aluminum GEC if GEC encased in concrete
- Functional grounded inverters include fuse grounded and transformerless inverters
- Functional grounded inverters can use either PV wire or USE-2 wire on source circuits
- Functional grounded inverters if fused in combiner only require fuses on one polarity
- PV connectors are polarized (positive and negative)
- Bypass diode failure usually decreases voltage by one-third of a module voltage
- 120/240 is called split phase or single phase and is used in residential wiring in North America
- 120V inverter has single pole ac disconnect/breaker and is mostly seen on off-grid systems
- No multiwire branch circuit sign is required with 120V inverter

- No more than six switches to turn off PV systems on a building (many inverters with one switch can be a single PV system)
- No disconnect on grounded conductor with exceptions for GFDI, AFCI, and maintenance
- PV source and output circuits must be separated from other circuits
- PV circuits must be polarized, marked, latching, and identified
- Load break rated manual disconnect within 10 feet of combiner required if current > 30A
- Isolating device can be a connector or a disconnect. Not required to interrupt current
- Isolating devices or disconnects need to be within 10 feet of equipment.
- Shading short edge of module typically kicks in all bypass diodes and bypasses module
- Interactive inverters do not need clamped breaker
- Bond rails to each other (UL 2703 is for racking systems)
- UL 1741 is listing for inverters
- UL 1703 is listing for PV modules
- Max dc disconnect height under normal circumstances is 6.5 feet
- PV system disconnect with utility disconnect in different places, need plaque or directory
- For lightning use lightning protection system and surge protection
- Add acid to water so acid doesn't splash
- EMT supports required at least every 10 feet
- No disconnect or fuse on solidly grounded conductor
- No fuse by disconnect required on functional grounded conductor
- Changing inverter make sure string sizing works with new inverter
- Always follow manufacturer's instructions
- Use torque wrench for installing lag bolts
- Solar system, HVAC and anything permanent is a dead load
- Designated safety person may not have other duties (distractions)
- No dc disconnect in bathroom (wet feet)
- Do not bond neutral in two places (dc disconnect and inverter GFCI)
- Person who puts on the lockout tagout removes it
- Insulation tester is megohmmeter or megger
- If disconnects not near one another use plaque or directory
- Battery bank sign should indicate grounded conductor and max voltage
- Equalization and cold temperature battery corrections increase max voltage

- Equalization only for flooded lead acid battery maintenance
- Equalization fixes stratification and lead sulfate crystal on lead plates
- Battery working spaces require illumination (lights)
- Battery working spaces lights must not be automatically controlled only
- OSHA = Occupational Safety and Health Administration
- Employer can train and certify employee to work heavy equipment
- Fiberglass ladder with aluminum steps is safest
- Painted wooden ladder is unacceptable (paint hides cracks)
- Three-phase, 4 wire delta (stinger) high leg is orange

TRIGONOMETRY: SOH CAH TOA

SOH sin = opposite/hypotenuse
CAH cos = adjacent/hypotenuse
TOA tan = opposite/adjacent

Figure 1.1 Right triangle

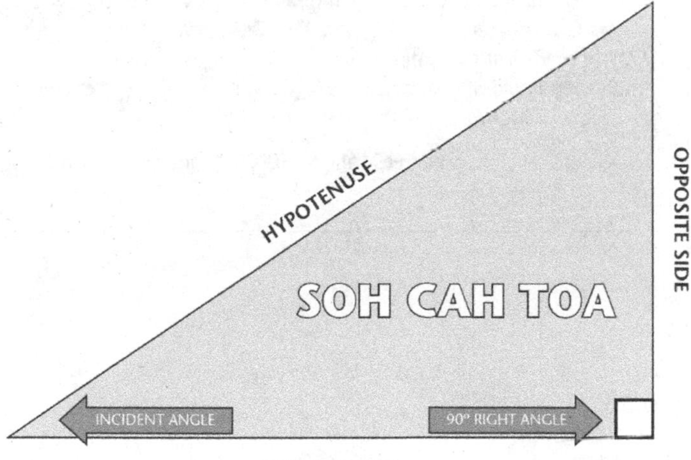

> **String sizing**
>
> PV source circuit string sizing calculations were done in detail in the first book in this series, *Solar Photovoltaic Basics*.

Calculator clicks (buttons that you push on a calculator) for determining maximum PV source circuit size without using paper:

Delta T × Temp Coef Voc in %/C = A
A%/100% = B
B + 1 = C
C × Voc at STC = Voc cold temp
1/X button
multiply by high inverter input voltage
Round-down for max modules in series

> **Temperature coefficients and units**
>
> A 1-degree change in degrees K is the same as a 1-degree change in degrees C. Kelvin starts at 273 below zero Celsius. When doing temperature coefficient calculations and you see degrees in K, just do the same as you would when it is degrees C. (C and K have almost the same phonetic sound anyway.)
>
> A 1.8-degree change in degrees Fahrenheit is the same as a 1-degree change in degrees Celsius or Kelvin.

Here is an example of using this method:

Given:

Low temp = −10 (Delta T is −35 and you can always do this in your head)
Temp Coef Voc = −0.33%/C
Voc of module = 40V
Max inverter input = 600V

35C × 0.33 %/C = 11.55% (you could multiply neg × neg = pos)
11.55%/100% = 0.1155
0.1155 + 1 = 1.1155
1.1155 × 40Voc at STC = 44.62Voc cold temp
1/X button gets you the inverse of 44.62, which is 0.022411
0.022411 × 600V = 13.4
13 modules in series max.

> This can be done without writing anything down. Practice 10 times and you should be able to do source circuit "string" sizing in 1 minute.

Calculator clicks for determining roof angle:

Example 4:12 slope roof

Enter:

4/12 = Inv Tan
Gives you 18.4 degrees
(inv tan is sometimes represented by $\tan^{-1}$)

Calculator clicks for the 120% rule to determine max inverter ac amps:

Enter:

Bus × 1.2 = −Main = /1.25 = inverter ac current

120% rule to solve for inverter power:

(((Busbar × 1.2)−Main) × 0.8) × grid voltage = max. inverter power

The National Electrical Code

Chapter 2

Every chapter in this book will focus on the National Electrical Code. The NEC is so important that it also deserves its own chapter at the beginning. Since it is an open-book exam, it is time to start getting close with your friend, the NEC. (Having a positive attitude about your NEC Codebook will help.)

As of the writing of this book, NABCEP has decided to let computer based testing examinees use an electronic version of the NEC that is on the computer at the testing center that you would be taking your test on. The first round of test takers with this new format had sufficient complaints about the navigation of the online NEC that NABCEP decided to allow people to bring in their own NEC Codebook. The computer based test (CBT) is also a take any time test. In the past, the NABCEP certification exams were only given in person twice per year. NABCEP will no longer be giving in-person exams. Be sure to check with NABCEP for details. They are easy to contact by phone or email, which can be found at www.nabcep.org.

The NEC, which is published by the National Fire Protection Association (NFPA), is published every 3 years. For instance, you may be taking the exam that is based on the 2017, 2020, or 2023 NEC. Usually, NABCEP will adopt the code one to one and a half years after the code is published. If you study the wrong version, you will probably miss zero to one more questions than you otherwise would have. The goal of the exam is not to fail people studying earlier versions of the Code.

Only about two states adopt the NEC as soon as it is published. This gives everyone else time to learn the NEC. Most of the solar installed in the USA is installed in states that adopt the NEC 3 years after it is first published. For example, in California, the 2017 NEC will not be used until 2020.

The NEC is officially called NFPA 70®: National Electrical Code. It is organized mainly into Chapters, Articles, Parts and Sections. Chapters 1 through 4 are

Table 2.1 Organization of the National Electrical Code

General for all installations	Chapter 1	General
	Chapter 2	Wiring and protection
	Chapter 3	Wiring methods and materials
	Chapter 4	Equipment for general use
Supplements or modifies Chapters 1–4	Chapter 5	Special occupancies
	Chapter 6	Special equipment (PV goes here)
	Chapter 7	Special conditions
Communications	Chapter 8	Communication systems
Tables	Chapter 9	Tables

general and apply to all electrical installations. Chapters 5, 6, and 7 are special and can modify Chapters 1 through 4. Chapter 8 is for communications and Chapter 9 is tables.

For example, Article 690.2 Definitions is in Chapter 6, Article 690 and Section 690.2. Articles, such as Article 690, can be divided into parts.

The version of the NEC that you will likely be using will be a paperback softbound copy of the NEC Code Book. It is recommended that you get the exact same book to study from.

Here are the different types of NEC books and accessories:

1. NEC Codebook (softbound): This is the book that NABCEP uses. It is recommended to study with this book. If you do not have the NEC Codebook, it is recommended that you order one now.
2. NEC Handbook: This has everything that the NEC Code Book has with added explanations and images. This is good to have, but is not what you will be using for the exam.
3. NEC on PDF: These are available and helpful, but not what you use on exam day. If you take the computer based exam, the electronic format that you are able to use is different than the PDF you can buy. The PDF that you can buy is easy to search and index. Once you pass the exam, the PDF version is easiest to use in my opinion.

4. Free NEC, available from www.nfpa.org, can only be used with a connection to the internet and does not currently have a search function. It is difficult to use, but some say easier to use than the electronic format you will have access to use at the testing center.
5. NEC books with tabs: Helpful, but probably not what you use on the exam. Check with NABCEP for the latest.

It is highly recommended that you get very used to using the paper NEC Codebook and learn how to use the index and the table of contents. There will be information on the exam that you did not expect to be there that is in the NEC. If you know how to use the book efficiently, you will be able to answer these questions. I recommend that the night before the exam, besides getting a good night's sleep, you should read the index and table of contents. If you try to do too much the night before, you will probably just stress yourself out, which is not a good test-taking strategy.

Some people will sleep with their NEC book under their pillow or practice turning pages, getting to Article 690 as fast as possible.

Coming up are some common places in the NEC that you should be able to look up quickly:

- Memorize the best you can.
- Look at the page in your NEC book if you have it with you.
- Consider flash cards (or an NEC tattoo).

The best way to use the remainder of this chapter is to read it once, knowing that you will not retain everything, and then to go back and study it more.

Starting off with some major categories in order of importance.

ARTICLE 690: PHOTOVOLTAIC SYSTEMS

Obviously, you should know how to look up everything here.

SECTION 705.12: INTERCONNECTIONS

Section 705.12 is almost a part of Article 690. In fact, in the 2011 NEC they moved material from 690 to 705.

We will cover Articles 690 and 705 in greater depth later in their own dedicated chapters. We will now cover the important code in numerical order.

ARTICLE 710: STANDALONE SYSTEMS

Another former part of Article 690, in the 2017 NEC, this material was moved from Section 690.10 to its own article. Since Article 710 and Section 690.10 both end with 10, it is convenient for memorization.

ARTICLE 706: ENERGY STORAGE SYSTEMS

They say this is the future, so good to get to know. Much of the material here is new. We still have Article 480 Storage batteries, but energy storage systems (ESS) can encompass more than just chemical batteries, 706 can include the electronics that are part of the system. Many of our batteries today come factory assembled with controls.

Now in numerical order (wording of definitions is shortened here for easiness of use).

CHAPTER 1: GENERAL

ARTICLE 100

Definitions

Following are a few important definitions written in plain English:

Accessible (equipment): **Not behind locked doors**, elevation or other means.

Readily accessible: Not requiring tools, ladders or climbing to gain access. **May be behind locked doors.**

Bonding: Connected for electrical continuity.

Multiwire branch circuit: Branch circuit with two or more ungrounded conductors and one neutral grounded conductor.

Control circuit: Circuit that controls performance of another circuit. (Can be used with 690.12 rapid shutdown for turning off power at combiner.)

Feeder: Conductors between primary and secondary distribution.

Field labeled: Equipment that was evaluated by a Field Evaluation Body and subsequently labeled by that body. For example, for a price UL can come to your site and compensate for lack of a factory listing of a product.

Grounded, solidly: Connected to ground without any resistance or impedance device. (Note: functional grounded inverters are not solidly grounded.)

Interrupting rating: The highest current that a device is intended to interrupt.

Photovoltaic (PV) system: The total components and subsystem that, in combination, convert solar energy into electric energy for connection to a utilization load.

Raceway: Channel for holding wires, cables or busbars, such as conduit.

Separately derived system: Electrical source having no direct connection to circuit conductors other than grounding and bonding. Examples are transformers.

Service: Conductors and equipment delivering (serving) electricity from utility.

110.14: Electrical Connections

Terminals and splices, which includes 110.14(C) Terminal temperature limitations.

110.21(B): marking/field applied hazard marking (for caution, warning or danger signs)

110.21(B) is referred to many times throughout the NEC, including in Articles 690 and 705. It recommends in an informational note to look to *Product Safety Signs and Labels ANSI Z535.4-2011* for guidance.

Labels should be permanently attached and not hand-written unless the marking is *"variable or subject to change."*

110.21(B) applies to PV systems in the following parts of the code. In these places, it says: **"Labels shall comply with 110.21(B)."**

690.13(B) label for when both sides of a disconnect may be energized in the open position label shall comply with 110.21(B).

Figure 2.1 Line and load label, courtesy of pvlabels.com

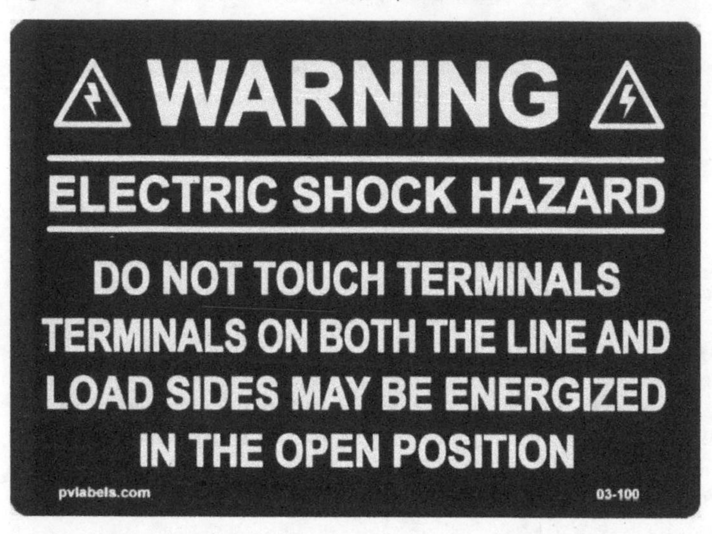

710.15(C) (formerly 690.10(C)) states that 120V standalone systems cannot have 120/240V ac multiwire branch circuits:

Figure 2.2 Stand-alone 120V inverter label, courtesy of pvlabels.com

Article 710 Standalone systems, which was relocated from Section 690.10 in the 2017 NEC. The reason we do not use multiwire branch circuits in this case is so that we do not overload the neutral. If we have a 120/240V split-phase panelboard and we bond line 1 to line 2 in order to modify the panelboard for a 120V system, then we would have line 1 and line 2 in phase and we would have double the current on the neutral. This is often done when using a 120V inverter rather than a 120/240V split-phase inverter.

705.10 (Formerly 690.56(B)) Facilities with utility and PV need a plaque or directory showing where the disconnect is if it is not at the same place as the utility disconnect.

705.12(B)(2)(3)(b) Requires a **110.21(B)** compliant sign when the 120% rule is invoked and the solar breaker is required to be on the opposite side of the busbar from the main breaker. This sign must be adjacent to the solar backfed breaker:

Figure 2.3 120% rule label, courtesy of pvlabels.com

When the people who write the NEC came up with **705.12(B)(2)(3)(b)**, they were making sure that we would get the NEC book when we take the NABCEP exam, because we cannot be expected to memorize **705.12(B)(2)(3)(b), which takes 26 key strokes!**

705.12(B)(2)(3)(c) gives us a way to get around the 120% rule in some cases. Read the sign and you will understand what **705.12(B)(2)(3)(c)** lets us do:

WARNING:

THIS EQUIPMENT FED BY MULTIPLE SOURCES.
TOTAL RATING OF ALL OVERCURRENT DEVICES,
EXCLUDING MAIN SUPPLY OVERCURRENT DEVICE,
SHALL NOT EXCEED AMPACITY OF BUSBAR

705.12(B)(2)(3)(c) is often used when combining inverters in a subpanel, sometimes referred to as an "ac-combiner." When we had to follow the 120% rule in the past with subpanels, we were very limited in our options and solar installers felt it was unfair. Now we can. For more information see Chapter 705.

Table 2.2 From Table 110.26(A)(1) © 2014 NFPA

Voltage to ground	Minimum distance		
	Condition 1	Condition 2	Condition 3
1 to 150V	3 feet	3 feet	3 feet
151 to 600V	3 feet	3.5 feet	4 feet
601 to 1000V	3 feet	4 feet	5 feet

Condition 1: Exposed live parts on one side and no live or grounded parts on the other side. Also, exposed live parts on both sides that are guarded and insulated.

Condition 2: Exposed live parts on one side and grounded on the other. Concrete materials are considered grounded.

Condition 3: Exposed live parts on both sides of working space.

Table 110.28: enclosure selection

This table has the various **NEMA enclosure types**, including NEMA 1, 2, 3, 3R, 3S, 3X, 3RX, 3SX, 4, 4X, 5, 6, 6P, 12, 12K and NEMA 13. Table 110.28 will show you what conditions the NEMA enclosures will handle, such as **rain**, **wind**, **dust**, **submersion**, etc. (Good to know 110.28 for NABCEP exams.)

CHAPTER 2: WIRING AND PROTECTION

ARTICLE 200: GROUNDED CONDUCTORS

200.6(A)(6)

6 AWG and smaller can be marked white for PV single conductor circuits (typically PV wire and USE-2 PV source circuits).

> With the new in 2017 NEC functional grounded definition, it would be extremely rare to have a white wire or a marked white wire in a PV system. The only reason would be having a solidly grounded array, such as with a direct well pump.
>
> It is recommended when converting an older fuse grounded inverter to a modern typical non-isolated (transformerless) inverter to apply 200.6(A)(6) in reverse and to mark the white wire with a black marking. This may be accepted by some AHJs and not by others. Non-isolated inverters are much safer than fuse grounded inverters, so the authors of the NEC do not want to do anything to discourage the use of safer inverters. Recall that also USE-2 wire, as well as PV wire may be used for all inverters now.

ARTICLE 240: OVERCURRENT PROTECTION

240.4(B): overcurrent protection device (OCPD) rated 800A or less

Round-up to the next standard OCPD size. This means that it is possible for a conductor to be rated less than the OCPD.

Once you understand it does not make sense, then you understand the code.

Rounding-up above the ampacity of the conductors for an OCPD is one of those things in the code that does not make sense! Since we want the OCPD to protect the conductor, we want the OCPD to open the circuit before the wire burns up.

Many people get stuck here. Once you understand it does not make sense, then you understand the code. We will point this out in a few different places in the code.

Why do they do this? When they came up with "ampacity" they left room for more current beyond what the conductor is rated, so perhaps a 10AWG wire that is rated for 40A can really handle 60A.

Another mechanism for safety is 240.4(D), "The small conductor rule."

You are always allowed to use a larger conductor and, most of the time, your calculations will give you a conductor with a greater ampacity than the OCPD.

240.4(D): small conductors (small conductor rule)

Minimum conductor size for OCPD for conductors 10AWG and smaller.

Table 2.3 Minimum OCPD for small conductors

OCPD	Minimum copper conductor size
30A	10AWG
20A	12AWG
15A	14AWG

It is important to remember this when sizing wires later in the book.

240.15: ungrounded conductors

Ungrounded conductors must have OCPD with some exceptions.

ARTICLE 250: GROUNDING AND BONDING

Grounding is connecting to earth and bonding is connecting everything together.

250.52: grounding electrodes

Look here for the different types of grounding electrode, such as a ground rod, metal pipe, metal building structure, concrete-encased electrode, ground ring, plate, etc.

250.66, 250.166, and 250.122

Sizing **grounding conductors** (if you look at these numbers, there is a pattern. If you remember the pattern, you can find the right section with three quick tries (pattern: **66, 166, 122**)).

250.66 is for sizing the ac grounding electrode conductor (**ac GEC**).

- Based on the largest ungrounded service entrance conductor in **Table 250.66**.

From Table 250.66, AC grounding electrode conductors (copper examples):

Table 2.4 From Table 250.66 ac GEC © 2014 NFPA

Size of largest ungrounded Service entrance conductor	Size of ac GEC
2 or smaller	8AWG
1 or 1/0	6AWG
2/0 or 3/0	4AWG

> Usually the ac grounding electrode conductor is already installed when adding PV to a building. Sometimes the AHJ will require a separate ac grounding electrode conductor when installing a supply-side connection.

250.166 is for sizing the dc grounding electrode conductor (**dc GEC**).

- Not smaller than the largest conductor and not smaller than 8AWG copper
- 250.166 does not have a table

> It is rare to have to install a dc grounding electrode conductor, since functional grounded inverters do not require a dc grounding electrode conductor, since they are not separately derived systems and they use the ac equipment grounding conductor for reference to detect ground faults. There is often confusion here among older installers who were taught to install a dc grounding electrode conductor way back (10 years ago). A dc only system would require a dc grounding electrode conductor.

250.122 is for sizing equipment grounding conductors (**ac EGC** and **dc EGC**).

- Based on OCPD (overcurrent protection device)
- If no OCPD as with 1 or 2 strings, then use 156% of Isc in place of OCPD in Table 250.122 rating as per 690.45 and 690.9(B). (This was changed from Isc in the 2014 NEC.)

From Table 250.122, Minimum size of equipment grounding conductor (EGC):

Table 2.5 From Table 250.122 ac and dc EGC © 2017 NFPA

Overcurrent device size	Copper EGC minimum size
15A	14
20A	12
60A	10
100A	8

Since you have the book, use it. There are often exceptions to rules.

CHAPTER 3: WIRING METHODS

TABLE 300.5: MINIMUM COVER REQUIREMENTS

Look up here how deep to bury conduit and direct burial cables. Good material for exam questions and real life.

SECTION 300.7: RACEWAYS EXPOSED TO DIFFERENT TEMPERATURES

300.7(A): sealing raceways

Raceways that have different portions that are exposed to different temperatures (such as a freezer) need to be sealed.

300.7(B): expansion, expansion-deflection and deflection fittings

Expansion of raceways (and even solar rails) have temperature coefficients for expansion, which you can calculate with methods similar to temperature coefficients for voltage. We have a practice test question later in the book and have done the calculations for an expansion fitting. Often, in the field, installers will use an approved flexible conduit at certain locations in order to allow for expansion.

ARTICLE 310: CONDUCTORS FOR GENERAL WIRING (WIRE SIZING – MOST POPULAR PAGES OF THE NEC)

Know how to use the following **Article 310 wire-sizing tables.** Later in this book, we will size wires with these tables.

> Tables 310.15(B)(2)(a), 310.15(B)(3)(a), 310.15(B)(16) and 310.15(B)(17) are perhaps the most used pages in the NEC. Many people copy these pages – so they do not get worn out and for convenience.
>
> 310.15(B)(3)(c) Ambient temperature adjustment for raceways or cables exposed to sunlight over rooftops was REMOVED in the 2017 NEC.

Table 310.15(B)(2)(a)

Ambient temperature correction factors. (Hot temperatures mean wire will heat up faster.) The same information in 310.15(B)(2)(a) is in **Table 690.31(A)** with very minor exceptions. The NEC tells us to use 690.31(A) rather than 310.15(B)(2)(a), but there will likely never be a difference in the outcome during your long and prosperous career.

Table 2.6 From Table 310.15(B)(2)(a) © 2017 NFPA

Ambient temperature	75C conductor	90C conductor
26–30C	1	1
31–35C	0.94	0.96
36–40C	0.88	0.91

Table 310.15(B)(2)(b)

Do not use Table 310.15(B)(2)(b), which is based on 40C ambient temperatures. Some people deface their NEC, so they do not actually use this page.

Table 310.15(B)(3)(a)

Adjustment factors for **more than three current-carrying conductors** (too many conductors in a pipe cannot dissipate heat efficiently).

From Table 310.15(B)(3)(a), Derating for more than three current carrying conductors in raceway or cable:

Table 2.7 From Table 310.15(B)(3)(a) © 2017 NFPA

Number of conductors	Derating factor
4–6	80
7–9	70
10–20	50
21–30	45
31–40	40

Table 310.15(B)(3)(c) REMOVED

Ambient temperature adjustment for **raceways or cables exposed to sunlight over rooftops**. If a raceway is installed less than 7/8" from a rooftop, then a 33C temperature adder should be added to the high ambient design temperature, however, you should not install a raceway less than 1" from a rooftop, so that there is space for debris to go under the raceway.

Table 310.15(B)(16)

Sizing conductors in raceway, cable or earth (everything that is not in free air).

From Table 310.15(B)(16), Allowable ampacity for conductors in conduit at 30C:

Table 2.8 From Table 310.15(B)(16) © 2017 NFPA

Size of wire	75C rated conductor	90C rated conductor
14AWG	20A	25A
12AWG	25A	30A
10AWG	35A	40A

Table 310.15(B)(17)

Sizing single insulated conductors in free air (if wire is in free air, it can dissipate heat more easily).

From Table 310.15(B)(17), Allowable ampacity for single insulated conductors in free air at 30C:

Table 2.9 From Table 310.15(B)(17) © 2017 NFPA

Size (AWG)	75C	90C
18AWG	–	18A
16AWG	–	24A
14AWG	30A	35A
12AWG	35A	40A
10AWG	50A	55A

> If you are having trouble understanding the terminology, such as **ampacity, OCPD, raceway, electrode, EGC, GEC, 10AWG**, etc., it is recommended that you study the first book in this series, *Solar PV Basics*, in order to make the most of your time and your educational experience.

ARTICLE 314: OUTLET, DEVICE, PULL AND JUNCTION BOXES; CONDUIT BODIES; FITTINGS; AND HANDHOLE ENCLOSURES

You may need to look up some rules regarding how much space you need in a junction box. Just be familiar with how to look up Article 314.

ARTICLES 320 THROUGH 362

Table 2.10 Articles for raceways and cables

Article	Raceway or cable articles used most often with PV installations
320	Armored cable: AC
330	Metal-clad cable: MC
334	Nonmetallic-sheathed cable: NM, NMC and NMS (romex)
338	Service entrance cable: SE and USE (USE-2)
340	Underground feeder and branch-circuit cable: UF
342	Intermediate metal conduit: IMC
344	Rigid metal conduit: RMC
348	Flexible metal conduit: FMC
350	Liquidtight flexible metal conduit: LFMC
352	Rigid polyvinyl chloride conduit: PVC
356	Liquidtight flexible nonmetallic conduit: LFNC
358	Electrical metallic tubing: EMT (most commonly used)

These articles are about different **types of raceway, conduit** and **cable**. Know how to use the NEC table of contents to find the correct wiring method.

MC has different meanings in the solar industry. MC cable in the NEC is metal-clad cable, which is a cable assembly commonly used for PV dc circuits inside a house. PV source and output circuits have to be in MC cable or a metal raceway (usually EMT) from the PV until the first readily accessible dc disconnect when within a building. The other type of MC is the MC connector, which is often the type of connector in between the PV modules. Multi Contact is a company that makes the MC connectors, so in this case MC is for a brand name of a connector. The most common type of module interconnect is MC-4 or MC-4 compatible.

CHAPTER 4: EQUIPMENT FOR GENERAL USE

ARTICLE 480: STORAGE BATTERIES

To remember this, think of **48**-volt battery bank for **48**0. Batteries are also covered in Article 706 Energy storage systems and many other places in the code, which will change rapidly over the next few code cycles as energy storage becomes mainstream, in order to bottle the sunlight captured by PV systems. A common question is: "Where should we look in the NEC when installing batteries?" and the answer is not so clear. When the 2017 NEC was written, some thought that the old lead-acid oriented Article 480 would be taken out of the Code, but here it remains … for now. Perhaps if lead-acid batteries are installed, we could look primarily to article 480 and when other technologies are used, we can look to 706. Often lithium batteries are more than just batteries, they often have electronics working with them in an enclosure that is only to be opened by the factory, making them more of an energy storage system than just a battery. Another clue we can use as to which article will apply to your installation is that Article 706 applies to energy storage systems over 50Vac and 60Vdc. That leaves less than 60Vdc for article 480.

480.2: definitions/nominal voltage

Nominal means "in name only" and it says in the code that nominal voltages are given for "convenient designation." For example, your car battery is 12V (2V per cell) nominal. When it is sitting it will be about 12.6V (2.1V per cell), when it is being charged it will be somewhere around 14.5V (2.4V per cell). Your car battery will be

12V (2V per cell) if it is somewhere in the neighborhood of halfway discharged, which hopefully never happens. It will also drop down in voltage for a split second when you are starting the car with an electric starter. Most car batteries are lead-acid batteries and have about the same voltage characteristics as other lead-acid batteries.

Nominal voltages from Informational Note in Section 480.2:

Table 2.11 Battery chemistry nominal voltages

Battery technology	Nominal voltage per cell
Lead-acid	2V
Alkali	1.2V
Lithium ion (Li-ion)	3.6 to 3.8V

> Informational note: informational notes in the code are good advice or tips and not rules that have to be followed in the code. The AHJ, however, can enforce whatever it sees fit to enforce.

Definition of sealed cell or battery: A sealed cell cannot accept addition of fluid for maintenance. A sealed lead-acid battery will have a valve to release pressure.

480.10(C): spaces about battery systems

For battery racks, there should be a minimum clearance of 1" between a cell container and any wall or structure on the side not requiring access for maintenance.

490.2: high-voltage definition

Over 1000V (for purposes of this article):

> Oftentimes we talk about medium-voltage ready inverters that are connected on the ac side to thousands of volts.

If you are confused about the different definitions of high, medium, and low voltages, do not worry. It is confusing for everyone. One thing is for sure, though, medium voltage is thousands of volts.

CHAPTER 5: SPECIAL OCCUPANCIES

If you are installing PV systems at a **gas station, hospital, trailer park, marina, aircraft hangar**, or in a **barn with animals**, you should check Chapter 5 for special requirements.

CHAPTER 6: SPECIAL EQUIPMENT

This is where PV falls, along with electric vehicles, industrial machinery, IT equipment, swimming pools, fuel cells, wind, and other special equipment.

ARTICLE 690: PHOTOVOLTAIC SYSTEMS

This is covered in Chapter 3 of this book.

ARTICLE 691: LARGE-SCALE PV ELECTRIC POWER PRODUCTION FACILITY

This article covers very large PV systems and is not the focus of any NABCEP exams. Article 691 is covered in more depth in another book in this series *Photovoltaic Systems and the National Electrical Code*. That being said, let us take a look at Article 691.

Article 691 was written because PV systems were the only large power plants that were being looked at by local electrical inspectors. Most large power plants are not subject to the NEC and are covered by the National Electrical Safety Code. These systems are usually manned, have big fences around them, and are only accessible by qualified people. It is not like a building where a kid might climb on the roof to get a frisbee. Much of the intent of 691 is to have a way to work around the requirements of Article 690 when under engineering supervision.

691 only applies to systems that meet all of these requirements:

- 5MWac and larger systems.
- Sole purpose of supplying utility power (no loads or net-metering).
- Accessible only to authorized personnel.
- Connection through medium or high voltage switchgear.
- Not on buildings.

Systems may be larger than 5MWac and are not required to invoke Article 691. 691 requirements are less safe in a way, but more safe because of special procedures to limit access.

691 may allow systems to forgo the arc-fault requirements of 690.

> The National Electric Safety Code (NESC) is the code that the utilities use. You can see that utility power lines go high in the sky. They are many thousands of volts and the conductors are often bare, with no insulation at all. Why does the bird sitting on the line not get bothered? It is not bothered unless it is completing the circuit. There are different rules for equipment on utility property, including power poles. Oftentimes people consider large "utility-scale" PV plants to be the property of a utility and not required to be NEC-compliant.

Figure 2.4 Bird on bare high voltage line, Wikimedia: http://commons.wikimedia.org/wiki/ File:BirdChaparalles.JPG

CHAPTER 7: SPECIAL CONDITIONS

There are articles in Chapter 7 that we will use on a regular basis in the solar industry and this is becoming more prevalent with the modernization of the grid. We will only cover the renewable energy relevant articles in Chapter 7.

ARTICLE 705: INTERCONNECTED POWER PRODUCTION SOURCES

This is covered in Chapter 4 of this book and includes how we connect to the grid.

ARTICLE 706: ENERGY STORAGE SYSTEMS (ESS)

706.1: scope

Article 706 applies to permanently installed energy storage systems over 50Vac or 60Vdc. (If there were an energy storage system under 50Vac, if would fit neither in 480 nor 706.)

706.2: definitions

Energy storage system (ESS): "One or more components assembled together capable of storing energy for use at a future time. ESS(s) can include, but are not limited to **batteries, capacitors**, and **kinetic energy devices (e.g., flywheels and compressed air**). These systems can have **ac or dc** output for utilization and can include inverters and converters to change stored energy into electrical energy."

Discussion: Article 706 covers modern energy storage devices, many which are in the state of development and will undoubtedly look different in a decade. Some say that super-capacitors are the holy grail of energy storage, if they could only figure out how to make them hold more energy. If you are studying for a NABCEP exam, they often do not cover recent changes. It has been said that as of 2019, nobody had yet seen a lithium battery question.

Self-contained energy storage system:

Components are assembled into a single energy storage unit.

Pre-engineered energy storage system (of matched components):

Components are field assembled from components supplied by a single entity.

Flow battery definition: "An energy storage component **similar to a fuel cell** that stores its active materials in the form of **two electrolytes** external to the reactor interface . When in use, the electrolytes are transferred between **reactor and storage tanks.**"

706.7: disconnecting means

Discussion: Disconnecting means shall be provided for all ungrounded conductors and be within line of sight and readily accessible or if controlled by remote actuation, the remote disconnect shall be capable of being locked in the open (off) position and the location of the controls shall be marked on the disconnecting means.

According to 706.7(D), the marking of the disconnecting means shall indicate:

1. nominal ESS voltage
2. maximum available short circuit current
3. clearing time or arc duration based on short circuit current and OCPD
4. date calculation was performed.

706.7(D): exception

Label not required if arc-flash label is applied.

Arc-flash labels are covered in NFPA 70E (NFPA 70 is the NEC, which is a different book to buy).

706.7(E)(3)

When fused disconnects are used, the line side of the disconnect should be connected towards the energy storage system.

This makes sense for a lot of renewable energy and it is always a good idea to put the line side of equipment towards where there could be more current if there were a short circuit.

706.10: ENERGY STORAGE SYSTEM LOCATIONS

706.10(A): ventilation

Ventilation is now only required to be appropriate with the energy storage technology. In previous version of the Code, ventilation was required for batteries and not all batteries require ventilation. This was because when lead-acid batteries are overcharged, they split H_2O into hydrogen and oxygen via electrolysis, which is a way to make explosive rocket fuel. When the inspectors of the past read the Code, they often would enforce ventilation for lithium batteries, which is no longer required, unless ventilation is in the manufacturer's instructions.

Anything in the manufacturer's instructions is always required according to 110.3(B).

706.10: directory

A plaque or directory shall be at each service entrance location and at all sources capable of being interconnected. If the facility is standalone, then the plaque or directory shall be readily visible on the exterior of the building and indicate the location of the standalone system disconnecting means.

706.10(C): spaces about ESS components

- Comply with 110.26 working spaces.
- For battery racks, a minimum of 1" between a cell and any wall or structure on the side not requiring maintenance or permitted to contact if free air is not less than 90% of the length.

706.20: CIRCUIT SIZING AND CURRENT

This is where we define currents, much like currents are defined in 690.8, which is where we define PV system currents. Some of this material was moved from Article 690 when we took most references to energy storage or batteries out of Article 690 in the 2017 NEC.

706.20(A)(1): nameplate-rated circuit current

The nameplate rated current of an ESS shall be the maximum rated current used for sizing wires for pre-engineered systems.

706.20(A)(2): inverter output circuit current

The inverter continuous output current is used. Note that standalone inverters often have greater surge current ratings that are not considered the rated current for wire sizing purposes.

706.20(A)(3): inverter input circuit current

The inverter input current shall be the continuous input current rating at the lowest voltage. Note that when input voltage is lowest, it takes the most current to get continuous power. Continuous current is a current that can be for 3 hours, whereas a surge current is less time.

706.20(B): conductor ampacity and overcurrent device ratings

Feeder circuit from ESS to wiring system serving loads shall not be less than the greater of:

1. Nameplate current
2. ESS overcurrent protective device

Occasionally, with normal wire sizing it is possible to have a wire with less ampacity than the overcurrent protection device due to rounding up of conductor ampacity. This is not the case here with ESSs.

706.21: OVERCURRENT PROTECTION

706.21(D): current limiting

Listed current limiting OCPD shall be installed for each ESS dc output, unless provided already in ESS device.

A current limiting overcurrent protection device, unlike regular overcurrent protection device is a device that can interrupt huge currents from short circuits, such as shorting a battery where all of the energy can come out a second and cause an explosion. A small OCPD that operates just above maximum rated current will not necessarily open a circuit of tens of thousands of amps.

706.21(E): fuses

If fuses are accessible to other than qualified persons, the fuse shall be able to be disconnected from each side. Pullout disconnects are acceptable to disconnect fuses.

706.23: CHARGE CONTROL

706.23(A): general

Charge control shall be required for ESS.

> Note that 690.72: self-regulated PV charge control does allow PV systems to forgo charge control as long as the PV system cannot charge the battery more than 3% in one hour and the voltage and current of the PV shall be matched to the battery. 690.72 is the only place where batteries are covered in 690 as of the 2017 NEC.

706.23(B): diversion charge controller

A diversion charge controller will regulate charging by diverting excess energy away from the charged ESS and to a diversion load or the grid. Common diversion loads are water pumps and electric heating elements in a hot water tank.

(1) Sole means of regulating charging

A diversion charge controller shall NOT be the sole means of regulating charge (there must be a backup way to prevent overcharging the ESS in case the diversion stops working).

706.23(C): charge controllers and DC-to-DC converters

This section talks about how the current can be determined by the device. It is interesting to note here that we are putting charge controllers and dc-to-dc converters in the same sentence, as if there was no difference between the two, which in a way is the truth, as far as maximum power point tracking charge controllers go. We have one dc voltage going in, usually a lower voltage and then another dc voltage going out with a charge controller, which is a dc-to-dc converter. Perhaps in the future, we will abandon the charge controller definition altogether and just call it a dc-to-dc converter. Unless if the charge controller was not maximum power point tracking, then it would not be converting the voltage, so an old style charge controller would not be a dc-to-dc converter.

Part III: electrochemical energy storage systems

Part III refers to batteries that are not part of a listed product, such as the old style lead–acid batteries of Article 480. Many of the modern batteries are part of a listed product that includes modern 21st century electronics. I would say that there could be a turf war brewing between those who write Article 480 and those who write 706 Part III, but I have no evidence. Who's side would you be on?

Part IV: flow battery energy storage systems

As we mentioned in 706.2 Definitions, a flow battery is similar to a fuel cell and has two tanks of electrolyte and a reactor, where the chemical reaction happens, whether charging or discharging.

An **electrolyte is a medium that provides the ion transport mechanism between the positive and negative electrodes of the cell.**

Flow battery electrolytes shall be identified by name and chemical classification with a sign wherever electrolyte can be accessed. Electrolyte shall be contained.

ARTICLE 710: STANDALONE SYSTEMS

Article 710 was, for the most part taken from Section 690.10 and energy storage was taken out of the domain of Article 690 in the 2017 NEC. This way we can condense wind, PV, and other systems that use energy storage into one article and save space. Article 710 is a half-page in the NEC.

710.15(A): supply outlet

The ac output of the inverter can be less than the calculated load. It does not have to supply everything turned on at once.

The ac output of the inverter must be at least as much as the largest load.

In practice, many off-grid rebels do not follow this rule and will have a big load that cannot be powered by the PV system and requires a generator.

710.15(C): single 120V supply

120V inverters are allowed to supply 120/240V equipment.

No multiwire branch circuits are allowed and warning signs are required.

Overcurrent device on output of the inverter must be less than the neutral bus of the service equipment.

Required label:

> WARNING:
>
> SINGLE 120V SUPPLY. DO NOT CONNECT MULTI-WIRE BRANCH CIRCUITS!

The reason for this sign is if we use a 12V inverter with 240V loads and convert the 120/240V equipment to 120V by bonding the line 1 and line 2 panelboard busbars together, then the neutral that balanced the formerly out of phase lines will now unbalance the currents on the neutral that was not designed for this extra current.

710.15(E): backfed circuit breakers

Plug-in, backfed circuit breakers shall be secured. This is because stand-alone inverters do not anti-island and if a breaker popped out, it would be energized!

Also, as for any backfed breaker, breakers marked line and load are not to be backfed.

ARTICLE 712: DIRECT CURRENT MICROGRIDS

Dc microgrids are covered here in Article 712 and ac microgrids are covered in Article 705, which has its own chapter in this book (4). Article 712 is about 1½ pages long and much of the information is repeated throughout the articles covered in this book.

So the big question is: What exactly is a dc microgrid? Can my flashlight be a dc microgrid or does it have to be bigger and the answer is, that there is no given minimum size for a dc microgrid, but it must be pretty obvious that your flashlight does *not* qualify, unless your flashlight could light up a small village.

712.2: definitions

Direct current microgrid (DC microgrid)

NEC: *A DC microgrid is a power distribution system consisting of more than one interconnected power source, supplying dc-dc converter(s), dc load(s), and/or ac load(s) powered by dc-ac inverter(s). A dc microgrid is typically not directly connected to an ac primary source of electricity, but some dc microgrids interconnect via one or more dc-ac bidirectional converters or dc-ac inverters.*

English: This means that to qualify as a dc microgrid, you need more than one dc power source and that there can be ac connections via inverters.

Dc power sources can include:

- ac-dc converters (rectifiers)
- bidirectional inverters
- PV
- wind
- energy storage
- fuel cells.

So, it could be argued that a small PV system with PV and batteries would have 2 dc power sources and could qualify as a dc microgrid, however when talking to the AHJ (or NABCEP), it is best not to argue.

ARTICLE 705

This is covered in Chapter 4 of this book.

CHAPTER 9: TABLES

Chapter 9 has various tables; those that are used most are for determining how many conductors fit in conduit and voltage drop tables.

TABLE 1: PERCENT OF CROSS-SECTION OF CONDUIT FOR CONDUCTORS AND CABLES

From Table 1, Percent of cross section of conduit for conductors and cables:

Table 2.12 From NEC Chapter 9 Table 1 © 2017 NFPA

Number of conductors or cables	Percent cross-sectional area you can fill
1	53
2	31
3 or more	40

This has to do with the geometry of how many circles you can fit in a circle.

TABLE 4: PERCENT AREA OF CONDUIT

These are the **internal dimensions of a cross-section of conduit.** Different types of conduit have different internal dimensions. We will just give the dimensions of most PV professionals' favorite conduit, EMT.

From Chapter 9, Table 4, Dimensions and percent area of conduit for electrical metallic tubing (EMT) (Article 358):

Table 2.13 From NEC Chapter 9 Table 4 © 2017 NFPA

Size	3 or more wires 40%	1 wire 53%	2 wires 31%	Internal diameter	Total area 100%
½ inch	0.122	0.161	0.094	0.622	0.304
¾ inch	0.213	0.283	0.165	0.824	0.533
1 inch	0.346	0.458	0.268	1.049	0.864
1¼ inch	0.598	0.793	0.464	1.38	1.496
1½ inch	0.814	1.079	0.631	1.61	2.036

See NEC Table 4 for more types of conduit.

TABLE 5: DIMENSIONS OF INSULATED CONDUCTORS

From Table 5, Dimensions of insulated conductors:

Table 2.14 From NEC Chapter 9 Table 5 © 2017 NFPA

Type of wire	Size	Area	Diameter
THWN-2	10AWG	0.0211 in^2	0.164 in
THWN-2	6AWG	0.0507 in^2	0.254 in
THWN-2	250kcmil	0.3970 in^2	0.711 in

With Chapter 9 Tables 1, 4 and 5 we find the total cross-sectional area of the conductors that we are going to put in conduit and then we use Table 4 to determine which size of conduit to use. It is always good to go a size higher for extra room. The information in Table 1 can be found in Table 4.

TABLE 8: CONDUCTOR PROPERTIES

Chapter 9Table 8 is the table that we use for voltage drop.

From Table 8, Conductor properties (simplified):

Table 2.15 From NEC Chapter 9 Table 8 © 2017 NFPA

Size	Strands	Resistance
10AWG copper	1	1.21 ohm/kFT
10AWG copper	7	1.24 ohm/kFT
6AWG copper	7	0.491 ohm/kFT
350kcmil	37	0.0367 ohm/kFT

This table is typically used for voltage drop calculations. The units are given in ohm/kFT, this would mean that a 10AWG conductor with a resistance of 1.24 ohm/kFT would have a resistance of 1.24 ohms with 1000 feet of the conductor. If your inverter was 500 feet away, you would need to calculate for

1000 feet of conductor, since electricity runs in circuits, which are round trips. Coated conductors in this table refer to conductors that have a coating, such as tin plating on the conductor, and does not refer to insulation.

We will do a brief example of a voltage drop calculation using this table.

Voltage drop example question

You have a PV array in a field and an inverter at a house 250 feet away. You are using 10AWG stranded copper wire. There is just one string and the voltage of the string operates at Vmp of 200V and an Imp of 5A. What is the percentage power loss due to dc voltage drop?

Answer

Voltage drop is calculated with Ohm's Law, which states:

$$\textbf{Vdrop} = I \times R$$

We know that current is $I = 5\text{A}$

We will solve for resistance by multiplying:

$$\text{ohm} / \text{kFT} \times \text{kFT} = \text{ohm}$$

The distance from the PV to the inverter is 250 feet, so there will be two wires to complete the circuit, therefore the wire length is:

$$250 \text{ feet} \times 2 \text{ directions} = 500 \text{ feet}$$
$$500 \text{ feet} / 1000 \text{ feet per kFT} = 0.5\text{kFT}$$
$$\text{Distance} = 0.5\text{kFT}$$

From looking up the properties of 10AWG stranded copper wire, we get 1.24 ohms/kFT, so:

$$\text{ohm} / \text{kFT} \times \text{kFT} = \text{ohm}$$
$$1.24 \text{ ohms} / \text{kFT} \times 0.5 \text{ kFT} = 0.62 \text{ ohm}$$
$$\text{Resistance} = 0.62 \text{ ohm}$$

Now back to Ohm's Law:

$$\textbf{Vdrop} = I \times R$$
$$\textbf{Vdrop = 5A} \times \textbf{0.62 ohm}$$
$$\textbf{Vdrop = 3.1 volts}$$

We now know that during peak sun conditions, we are losing 3.1 volts on the wire and if we measured the voltage at the array, it would be 3.1 volts higher than at the inverter.

To figure out our losses as a percentage we will divide voltage drop by system voltage and then turn that into a percentage:

(Vdrop / system voltage) × 100% = voltage drop percentage

(3.1V / 200V) × 100% = 1.55% voltage drop

Since voltage multiplied by current is power, then voltage drop percentage is the same as power loss, so we are losing 1.55% of our power in this case.

Answer to voltage drop question: 1.55% power loss at STC

There have been many test takers who have studied voltage drop and were surprised to find not one voltage drop question on the exam. The reason for this is, perhaps, that voltage drop is an efficiency calculation and not related to safety with PV systems. Another reason is that with constantly changing irradiance and temperatures, the data from which we do calculations change. It is unusual for an array to operate at STC and if we were using STC current, we would realize that most of the time during the morning, day, and evening that the current is well below STC current. The system in this example would likely be operating below 1% losses much of the time.

AC inverter voltage drop

It is interesting to note that since the direction of power is from the inverter to the tie-in point, that the voltage at the inverter will be higher at the connection. If the conductor is sized too small on the ac side, then the inverter can turn off with too much current, which would lead to increased voltage drop and can make an inverter anti-island (automatically disconnect from the grid).

Doing voltage drop calculations is very important in your career as a solar PV professional; however, it is not recommended you stay up late the night before an exam studying voltage drop. Voltage drop is not known to be a big part of the exam. You should study voltage drop more the night after you take your exam!

Onward!

Article 690: photovoltaic systems

Chapter 3

In this chapter, we will cover the important parts of Article 690 that are not entry-level PV topics. Article 690 covers PV, although the rest of the NEC applies to PV installations. Article 690 is one of the fastest changing parts of the NEC due to the ever growing popularity and rapid advancement of PV technology.

Article 690 is divided into parts:

- **I.** General.
- **II.** Circuit requirements.
- **III.** Disconnecting means.
- **IV.** Wiring methods.
- **V.** Grounding.
- **VI.** Marking.
- **VII.** Connection to other sources.
- **VIII.** Energy storage systems.

Just because the NEC is organized into parts does not mean that it is perfectly organized. For instance, there are plenty of requirements for PV system marking that are not in Part VI Marking. Just like Article 480 is for batteries, there is also Part VII Storage batteries in Article 690, which just tells you that energy storage systems must be installed in accordance with Article 706. Will the NABCEP exams have Article 480 questions on the exam? Perhaps! The NABCEP exams are not easy. Let the "Parts" be a resource, but do not rely on them. Rely on your skills to navigate the NEC, which you will gain with effort, experience, and from studying from this book.

Article 690.2 definitions were covered in detail in the first book in this series, *Solar Photovoltaic Basics*, and can be read verbatim in the NEC.

690.4: GENERAL REQUIREMENTS

690.4(D): MULTIPLE PV SYSTEMS

- If there are multiple PV systems located remotely from each other, then there shall be a directory or plaque at each PV system disconnecting means.
- Why all of the directories? So emergency personnel do not turn off one switch and think they turned everything off, like in most houses.

Figure 3.1 Plaque showing different disconnect locations, courtesy of pvlabels.com

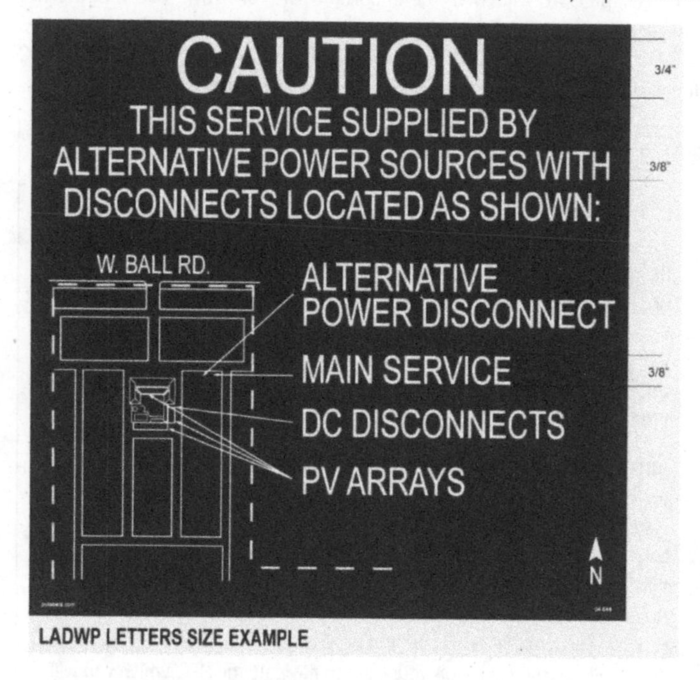

690.7: MAXIMUM VOLTAGE

The maximum voltage shall be the highest voltage between any two circuit conductors OR any conductor and ground.

Table 3.1 PV system dc circuit maximum voltages

Application	Max. voltage
1 and 2 family dwellings	600V dc
3 family dwellings and up	1000V dc
Not on buildings	Over 1500V dc
	Equipment must be listed for use
	Not required to conform to Article 490 Parts II and III
	Different rules
	Big deal

> There are many different definitions of high, medium, and low voltage. In many instances in the solar industry, we read about medium-voltage ready inverters, which connect on the ac side through a transformer at thousands of volts. You would be wise to know that there are many different definitions.

690.7(A): PV SOURCE AND OUTPUT CIRCUITS

There are three methods to choose from.

690.7(A)(1): instructions in listing or labeling of module

This is the way that is typically done using the low ASHRAE temperature from solarabcs.org and the module manufacturers' temperature coefficient of Voc.

In the *Solar Photovoltaic Basics* book, we went into detail about the maximum system voltage calculations and had an entire section on it. We will cover temperature voltage calculations again in the sample exam in this book, since it is very important.

690.7(A)(2): crystalline and multicrystalline modules

Table 690.7(A) can be used to cold temperature correct for the number of modules in series. This table is more conservative, so most designers prefer to do the

690.7(A)(1) method, which can also be used for crystalline and multicrystalline modules.

Table 3.2 NEC Table 690.7(A) Voltage correction factors for crystalline and multicrystalline silicon modules

Ambient temperature (C)	Factor
9 to 5	1.08
4 to 0	1.10
-1 to -5	1.12
-6 to -10	1.14
-11 to -15	1.16

690.7(A)(3): PV systems 100kW or larger

For systems with a generating capacity (ac output) of 100kW or more, under engineering supervision may use a licensed professional engineer (PE) to do the calculations and potentially get to put more modules in series.

Many installers have made the mistake of listing the inverter maximum input voltage as the maximum system voltage listed on the dc disconnect label. **The PV source and output circuit maximum system voltage is dependent on the PV, number of modules in series, and the cold temperature.**

690.7(B): DC-to-DC converter source and output circuits

690.7(B)(1): single DC-to-DC converter

Circuits connected to a single dc-to-dc converter, the maximum voltage shall be the maximum output of the dc-to-dc converter.

690.7(B)(2): two or more series connected DC-to-DC converters

Maximum voltage determined in accordance with instructions. If there are no instructions for series connected dc-to-dc converters, then the sum of the individual converters in series shall be used.

690.7(C): bipolar source and output circuits

Bipolar arrays are the dc version of 120/240V ac split-phase power. In a bipolar system, we have a positively grounded subarray and a negatively grounded subarray. Since there are two separate subarrays, we can refer to the voltage as "voltage to ground" as long as the conductors of the different arrays are kept separate in separate raceways and each subarray is functionally grounded. In effect, a 1500V system can get the efficiencies of a 3000V system. (Doubling voltage can increase efficiencies by four times.)

690.8: CIRCUIT SIZING AND CURRENT

Here we define the current that we use for sizing wires in (A) and we tell you the required ampacity in (B) with one of those 125% correction factors.

690.8(A): calculation of maximum circuit current

690.8(A)(1) through 690.8(A)(6)

These define current. This is easy for normal electrical devices that take current from the utility, but with solar, we have some interesting ways that this current can be more, such as we will read next in 690.8(A)(1).

690.8(A)(1): photovoltaic source circuit currents

There are two ways of calculating PV source circuit currents; 690.8(A)(1)(1) is the usual way that has been done in the past and 690.8(A)(1)(2) Is the new way for bigger projects.

690.8(A)(1)(1): USE Isc x 1.25 TO CALCULATE MAXIMUM CIRCUIT CURRENT

In the NEC, it is confusing because it says parallel connected when talking about PV source circuits. We always think of PV source circuits as series connected. The parallel is because in some large utility-scale projects, there are wiring harnesses that are paralleling some modules together to make a source circuit. I recommend ignoring this confusing "parallel" and just multiply Isc x 1.25 to get maximum circuit current.

690.8(A)(1)(2) for PV systems with a generating capacity (ac output) of 100kW or greater, a licensed professional electrical engineer can do calculations based on an industry standard method. This value cannot be less than 70% of the 690.8(A)(1)(1) method above. 70% of 125% of Isc is equal to 87.5% of Isc. (I have not heard of NABCEP covering 690.8(A)(1)(2).)

690.8(A)(2): photovoltaic output circuits

PV output circuit currents are the sum of the parallel connected PV source circuit currents.

> For PV source and output circuits, there are extra things that we have to do to make sure our conductors have enough ampacity. This is because, with PV, we can operate above the STC at which the PV was rated with increased irradiance beyond 1000 watts per square meter.

Figure 3.2 PV source and output circuit currents for conductor sizing

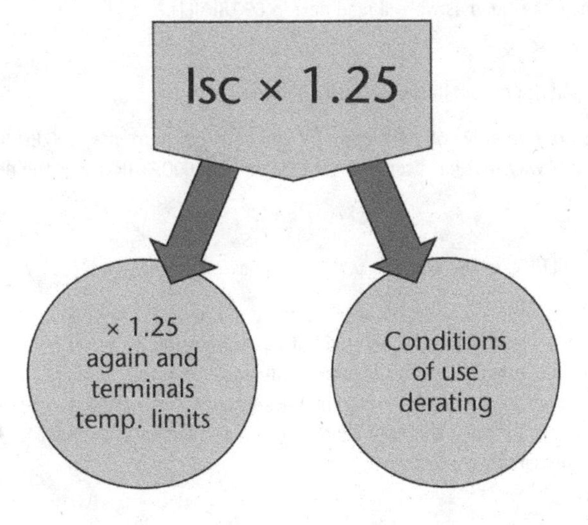

PV source and output circuit conductor sizing

1. Isc × 1.56 and check for the column of the temperature of the terminals or the conductor, whichever is less.
2. Isc × 1.25 and then derate for conditions of use, which are:
   a. ambient temperature derating
   b. temperature adders for conduit on a rooftop in sunlight
   c. more than three current-carrying conductors in a raceway.
3. Check that conductor is large enough for the OCPD in 240.4 and that the temperature corrected ampacity of the conductor does not exceed the OCPD.

We will check according to all three methods and then choose the largest conductor. Many PV professionals make the mistake of multiplying Isc × 1.56 and then derating for conditions of use, which is overkill. We do the calculations separately and then choose the largest conductor.

Another great mistake people make is using 1.56 for anything else but PV source and output circuits. No 1.56 on the inverter output!

We are just getting warmed up and will go over this in more detail later in the book with plenty of example calculations.

PV connected dc circuits have different definitions for current:

- Maximum power current = Imp — From PV Mfg.
- Short circuit current = Isc — From PV Mfg.
- Maximum current = Isc × 1.25 — 690.8(A)(1)
- Required ampacity for continuous current = Isc × 1.56 — 690.8(B)(1)
  - 1.25 × 1.25 is 1.56

The reason that we have this extra 1.25 factor for PV source and output circuits is because whatever it says on the PV label about how it was tested, it can operate above that current when it is brighter than STC. Other circuits are current-limited by electronics or loads.

690.8(A)(3)u: inverter output circuit crrent

The continuous output current shall be the maximum current.

Note that we do not consider surge currents here. For most interactive inverters, it is going to be watts divided by volts, however, some future-proof inverters can make reactive power, which can lead to greater currents, by changing the power factor. Power factor is calculated when voltage and current do not peak at the same instance, unless there is perfect power factor of 1. Less than 1 is out of phase and sometimes beneficial.

690.8(A)(4): standalone inverter input circuit current

The maximum current is the standalone inverter continuous current at the lowest input voltage.

At a lower voltage, it takes more current to get the same amount of power.

690.8(A)(5): DC-to-DC converter source circuit current

Maximum current shall be the continuous output current rating. DC-to-DC converters can control current.

690.8(A)(6): DC-to-DC converter output circuit current

Maximum current is the sum of the parallel connected dc-to-dc converter source circuit currents.

690.8(B): conductor ampacity

PV circuits are continuous (3 hours or more of operation) and **shall be sized to carry not less than 690.8(B)(1) or 690.8(B)(2), whichever is greater**. (If protected by electronic overcurrent protective device, which is rare, then there is another method.) This goes in line with how conductors are sized throughout the NEC.

690.8(B)(1): before application of adjustment and correction factors

Here we just take the maximum circuit currents that we defined in 690.8(A)(1) through 690.8(A)(6) and multiply by 1.25. This is the 1.25 for continuous currents. Additionally here, we will use the column in Tables 310.15(B)(16) and 310.15(B)(17) that correspond to the temperature rating of the conductor or terminals that the conductor is attached to, whichever is the lower temperature rating. This usually means the 75C column for the terminals.

690.8(B)(2): after application of adjustment and correction factors

Here we take the maximum circuit currents that we defined in 690.8(A)(1) through 690.8(A)(6) and derate them for conditions of use, which means we will lessen the ampacity of the wire based on Table 690.31(A) correction factors, which is almost the same as 310.15(B)(2)(a) and Table 310.15(B)(3)(a): adjustment factors for more than three current-carrying conductors. There is no accounting for continuous current or terminals in this step.

> Take note that as of the 2017 NEC Table 310.15(B)(3)(c): raceways and cables exposed to sunlight on rooftops was removed. This means less derating and more ampacity for these conductors.

690.9: OVERCURRENT PROTECTION

Overcurrent devices are not required if circuits have sufficient ampacity for highest available current. This often means that when you have one or two strings combined, that you do not need overcurrent protection. Other sources that may not have enough currents to require overcurrent protection may be interactive inverters and dc-to-dc converters. Usually three or more strings combined requires overcurrent protection.

When connected to sources of higher currents, such as the utility, batteries, and multiple parallel strings, then the protection shall be at the higher current source. In the case of an interactive inverter, the inverter cannot produce enough current to overheat a properly sized wire, since it is current limited, however, there needs to be overcurrent protection at the interconnection point, i.e., backfeed breaker.

> The unusual thing about PV and overcurrents is the lack of overcurrents, since PV is current-limited. The short-circuit current of a PV module is only about 7% more than the operating current of the PV in peak sun.

That means our overcurrent protection devices (OCPDs) will not open circuits in many cases, because of lack of overcurrents.

Where we see enough currents to open a circuit OCPD on a PV system:

- ac side when connected to utility
- when connected to batteries or other energy storage systems
- when multiple circuits backfeed through a combiner
- ground-fault fuse during ground fault on grounded inverter.

It is interesting to note that we could short circuit a megawatt of PV at the main inverter dc disconnect and not ever blow a fuse in any of the combiners for 20 years! This is because all of the fuses are sized for at least 156% of Isc, which will never happen.

The reason we have fuses in a combiner is for backwards currents when multiple circuits backfeed down a short on a single string.

Dc-to-dc converters also can limit current with smart electronics and the currents from an interactive inverter are limited to the nameplate output of the inverter.

690.9(B): overcurrent device ratings

OCPDs should be rated at 125% of maximum currents.

Maximum currents of PV source and PV output circuits are by definition 125% of Isc, therefore the OCPD of these circuits should be 156% of Isc. This means that the **fuses in the combiner are to be calculated by multiplying Isc × 1.56**.

Most installers know that two strings or fewer when combined do not need fuses in the combiner. This comes from the 690.9(A) exception (2):

> *The short-circuit currents from all sources do not exceed the ampacity of the conductors and the maximum overcurrent device size rating specified for the PV module or dc-to-dc converter.*

Then OCPD is not required, since it would never have enough current to open an OCPD if it were there.

Other circuit OCPDs are calculated much like circuits elsewhere in the code, by taking the maximum continuous current and multiplying by 1.25 and rounding up to the next common fuse size. On the ac side of an interactive inverter, the overcurrents that would open a circuit OCPD would be the current coming from the utility.

> Exceptions to the rule: Over 800A and for taps, we do not round-up to the next common OCPD size, to be extra safe.

690.9 also includes a few other methods, which include not having to use the 125% correction factor for overcurrent devices, which are rated for continuous operation at 100% of its rating and for adjustable electronic overcurrent protective devices. These other methods are rare, since thermally activated overcurrent protection devices are less expensive. Typical thermally activated devices, since they can heat up, need that extra 125%.

690.10: STANDALONE SYSTEMS

690.10 was moved to Article 710 Standalone systems in the 2017 NEC and is covered on page 36 in this book.

690.11: DC ARC-FAULT PROTECTION

This is required for dc circuits over 80Vdc and will detect the arc fault, interrupt the fault (open the circuit) and indicate there was a fault. This was new in the 2011 NEC.

690.11 Exception: If a PV system is not on or in a building, then the PV and dc-to-dc converter output circuits (not source circuits) that are direct buried or installed in metal raceways or enclosed metallic cable trays, then dc arc fault protection is not required for these output circuits.

Another way around 690.11 is by invoking Article 691 Large-scale PV, which, among other things, has to be at least 5MWac. 691 is covered on page 29 of this book.

690.12: RAPID SHUTDOWN OF PV SYSTEMS ON BUILDINGS

690.12 has had big changes in every Code cycle since its inception in the 2014 NEC. To make things more unique, some of the changes, most commonly known as module level shutdown, in the 2017 NEC do not take effect until 2019. It is also interesting to note that, in the 2017 NEC, the definition of a PV system has changed to not include energy storage or loads, so rapid shutdown is not required on energy storage system circuits or loads; it is PV system specific!

690.12(A): CONTROLLED CONDUCTORS

Rapid shutdown applies to **ac and dc PV system circuits on buildings.**

690.12(B): CONTROLLED LIMITS

Array boundary is defined as 1 foot from the array and applies to 690.12(B)(1) and (2).

690.12(B)(1): outside the array boundary (30V and 30 seconds)

Conductors outside of the array boundary (or 3 feet from the point of entry of a building) after rapid shutdown initiation shall be not more than 30V within 30 seconds.

Note that the 3 feet from point of entry is so that conductors are allowed to go immediately into the building and be shutdown inside of the building, which would be very difficult if you only had 1 foot to work with.

690.12(B)(1) for the most part means that you cannot even have the voltage on a single typical module 1 foot from the array.

690.12(B)(2): inside the array boundary (3 options)

Takes effect January 1, 2019.

690.12(B)(2)(1): listed rapid shutdown array

As of the writing of this book, there is no listing for rapid shutdown standard. It is likely that the standard will include module level shutdown.

690.12(B)(2)(2): MODULE LEVEL SHUTDOWN (80V 30 SECONDS)

Conductors inside the array boundary (or 3 feet from penetration of the building) must be able to go down to 80V within 30 seconds of rapid shutdown initiation.

The reason the writers of the Code came up with 80V is because most modules when corrected for cold temperatures stay under 80V, yet are over 40V, so a single module PV source circuit is your only option.

Most all compliance with 690.12(B)(2) will be using module level shutdown with microinverters, dc-to-dc converters and other devices that accomplish the same thing.

690.12(B)(2)(3): NO EXPOSED WIRING OR METAL PARTS

Arrays with no exposed wiring methods and no exposed conductive parts that are at least 8 feet away from exposed grounded conductive parts or ground shall be acceptable. BIPV is often made without metal frames and would be acceptable. Additionally, if someone came up with a way to mount PV without any exposed metal parts, such as a frameless module and a composite rack, it could be a way to get around module level shutdown.

This method was put in place as to make BIPV not impossible.

690.12(C): INITIATION DEVICE (MOST INITIATION DEVICES ARE A CIRCUIT BREAKER)

Initiation device shall indicate whether on or off and off means that rapid shutdown has been initiated.

Initiation devices on 1 and 2 family dwellings shall be in a readily accessible location outside the building.

Types of initiation device:

690.12(C)(1) Service disconnecting means

690.12(C)(2) PV system disconnecting means (usually backfed breaker)

690.12(C)(3) Readily accessible switch that indicates on or off (usually for buildings with battery backup)

690.12(D) Equipment

Rapid shutdown equipment that is not a listed disconnect, circuit breaker, or control switch shall be listed for rapid shutdown.

Note: For rapid shutdown signs, see 690.56(C) on page 67.

690.13 THROUGH 690.15: DISCONNECTING MEANS (THIS IS ALSO CALLED PART III)

690.13 Means shall be provided to disconnect PV systems from other systems, such as energy storage systems, premises wiring, and utilization equipment (loads).

690.13(A): LOCATION (READILY ACCESSIBLE)

Disconnecting means should be readily accessible.

690.13(B): MARKING

Must be marked as "PV SYSTEM DISCONNECT,"

If line and load may be energized in the open position, then the label must say this:

WARNING

ELECTRIC SHOCK HAZARD
TERMINALS ON THE LINE AND LOAD SIDES
MAY BE ENERGIZED IN THE OPEN POSITION

Note that interactive inverters often contain capacitors and electricians should be aware that even after the system is shutdown, there may still be energized conductors on the dc side of the inverter, even if rapid shutdown is in place. Rapid shutdown can still have 30V on the controlled conductors after 30 seconds, but still may have hundreds of volts for less than 30 seconds. Rapid shutdown does not always mean this "line and load" sign is not needed. Typically, ac sides of interactive inverters are going to anti-island immediately and do not require the "line and load" sign.

690.13(C): SUITABLE FOR USE (AS SERVICE EQUIPMENT FOR SUPPLY-SIDE CONNECTIONS)

Supply-side connected disconnecting means shall be suitable for use as service equipment (new in 2017 NEC).

60.13(D): MAXIMUM NUMBER OF DISCONNECTS

Each PV system disconnecting means shall have no more than six switches or six sets of circuit breakers. A single disconnect can be for the combined output of more than one inverter.

This does not limit the number of PV systems connected to a service.

690.15 DISCONNECTION OF PHOTOVOLTAIC EQUIPMENT

Isolating devices shall be provided to isolate:

- PV modules
- ac PV modules
- fuses
- inverters
- charge controllers.

A disconnecting means may be used in place of an isolating device.

For circuits over 30A a disconnecting means must be used for isolation.

Q: What is the difference of an isolating device and a disconnecting means?

A: An isolating device is not required to disconnect under load and not required to simultaneously disconnect. Examples of non-load-break-rated isolating devices are module connectors and fuse holders. Disconnecting means will disconnect non-grounded conductors simultaneously and under load.

690.15(A): LOCATION

- Isolating devices or disconnecting means shall be within the equipment or within sight and within 10 feet of the equipment.
- Disconnecting means may be remotely operated from within 10 feet of the equipment

690.15(C): ISOLATING DEVICE

An isolating device can be:

1. connector
2. finger safe fuse holder
3. isolating switch that requires a tool
4. isolating device listed for intended application.

Isolating devices shall be marked "do not disconnect under load" or "not for current interrupting" unless it is rated for disconnect under load.

Some microinverters and dc-to-dc converters are allowed to be disconnected under load with the PV module connectors.

690.15(C): EQUIPMENT DISCONNECTING MEANS

Must simultaneously disconnect conductors that are not solidly grounded under load.

690.31: WIRING METHODS PERMITTED

690.31(A): METHODS PERMITTED

We can use methods permitted in other parts of the NEC and methods specifically for PV.

If PV source and output circuits operate at a voltage over 30V in readily accessible locations, they shall be guarded or installed in MC cable or raceway.

TABLE 690.31(A): CORRECTION FACTORS

Table 690.31(A) is almost exactly the same as Table 310.15(B)(2)(A). These are tables that correct ampacity (ability to carry current) for ambient temperatures above 30C, so when your PV is outside in most places, you will have ambient temperatures above 30C. I would say that, 99% of the time, the tables are completely interchangeable and it is a waste of paper to put this table in the NEC. I have heard that it may leave the NEC in the next cycle. Table 690.31(A) does, however, have a column for 105C rated wire, which is uncommon.

690.31(B): IDENTIFICATION AND GROUPING

PV source and output circuits **(dc PV) should not be put in the same places as other non-PV circuits or inverter ac circuits** unless they are separated by a partition.

PV circuits should also be marked and identified.

When different PV systems are in the same location, they can be grouped using cable ties every 6 feet.

Identification can be done by color coding, marking tape, tagging, etc.

690.31(C): SINGLE-CONDUCTOR CABLE

USE-2 and PV wire may be used for exposed single conductor cable in outdoor locations. In the past PV wire had to be used for ungrounded inverters. This is no longer the case. We also no longer call them ungrounded inverters.

PV circuits may be put in cable trays even if the wire is not specifically rated for cable tray.

690.31(D): MULTICONDUCTOR CABLE

Multiconductor cable identified for use is okay for outdoor locations if secured every 6 feet (think microinverters).

690.31(G): PV SYSTEM DC CIRCUITS ON OR IN A BUILDING

DC source and output circuits and dc-to-dc converter source and output circuits in a building must be in a metal raceway or MC cable up until the first readily accessible dc disconnect. This is why most people use EMT. This requirement is not required on the outside of buildings or for ac circuits.

690.31(G)(3): marking and labeling

PV power source conductors must be marked with the wording:

WARNING PHOTOVOLTAIC POWER SOURCE

at exposed raceways, covers, enclosures, etc.

Labels to have reflective white letters at least ⅜" in height on a red background.

Labels at least every 10 feet.

690.31(I): BIPOLAR PV SYSTEMS

If the voltage between the separate "monopole subarrays" exceeds the rating of the equipment, then the **wiring for the separate subarrays must be installed in different raceways** and kept separate until the inverter.

Bipolar systems are essentially two different arrays, one positively grounded and the other negatively grounded. When both arrays connect at the inverter, the voltage is double of each individual array and the efficiencies increase. This is the dc version of the electricity going to an American house at 120/240V ac split-phase.

690.33: CONNECTORS

These must be:

- polarized
- guarded

- latching or locking
- require a tool for opening unless load-break-rated
- **marked "do not disconnect under load" or "not for current interrupting" unless rated for current interrupting.**

690.34: ACCESS TO BOXES

If a junction or other box is behind a module, the module should be easy to remove.

PART V: GROUNDING AND BONDING (690.35: UNGROUNDED PV SYSTEMS REMOVED IN 2017 NEC) 690.41: SYSTEM GROUNDING

System grounding refers to systems that have an intentionally current-carrying grounded conductor (white wire/neutral) that is bonded to and operates at the same voltage as ground. (In much of the world outside the US, the grounded conductor is blue.)

AC system grounding is done at the main service panel. This is usually done by connecting the grounded conductors (neutrals) to the same busbar as the grounding conductors or through a bonding jumper (not a fuse).

System grounding can only be done in one place. If the grounded conductor contacts ground in another place, then it would be a ground fault. There would be parallel paths for current to run through if there were two places where the white wire touches the green (or bare) wire and current would flow on the green (or bare) wire. Current should never flow on the green or bare wire.

On different sides of a transformer, there are separately derived systems and so system grounding would have to be done separately on each system.

With the 2017 NEC came a new category that encompasses most inverters installed today called "functional grounded." These systems are neither ungrounded nor solidly grounded. With these systems, we have no white grounded conductor, no requirement for PV wire, only a single polarity needs fusing when fusing is required, and both poles of a dc disconnect are required to open. The most common inverter used today is a functional grounded non-isolated inverter, which was formerly categorized as an ungrounded inverter in the 2014 NEC.

690.43: EQUIPMENT GROUNDING AND BONDING

Exposed non-current-carrying metal parts of PV module frames, electrical equipment and enclosures shall be grounded.

Listed, labeled and identified devices can be used for grounding (many racking systems are listed to UL 2703).

Equipment grounding is bonding equipment together so that nobody gets shocked by touching two different pieces of metal.

Every system, even if it is not an electrical system, is required to have equipment grounding if there is metal that could become energized by a stray wire. Even separately derived systems should have the equipment at the same voltage as other pieces of equipment of other systems.

A metal roof or a metal plumbing pipe or a steel frame of a building should be grounded.

We never want a loose wire to accidentally connect with metal and have the metal that is accidentally energized to be at a different voltage than something else in the building that someone could accidentally touch.

Oftentimes the equipment grounding conductor (EGC) is 6AWG or 10AWG underneath the PV modules. If bare copper is exposed and subject to damage, the smallest it can be is 6AWG. Some authorities having jurisdiction (AHJs) require a 6AWG EGC underneath the array, while others allow a 10AWG under the array. Oftentimes it is on the west coast of the USA where the wind speeds are not as high and the ice does not build up on the roof where they allow 10AWG. When the EGC is in conduit for arrays, often the wire gauge can be reduced to as small as a 14AWG conductor according to Table 250.122.

690.45: SIZE OF EQUIPMENT GROUNDING CONDUCTOR

Equipment grounding conductor is sized with Table 250.122 and is not smaller than 14AWG.

690.47: GROUNDING ELECTRODE SYSTEM

690.47(A): BUILDINGS OR STRUCTURES SUPPORTING A PV ARRAY

Shall have a grounding electrode system installed (they almost always do before the solar installer gets there).

PV equipment grounding conductors shall be connected to the grounding electrode system (this is done through the inverter equipment grounding connection in most inverters, back to a panelboard, just like any load is grounded).

690.47(B): ADDITIONAL AUXILIARY ELECTRODES FOR ARRAY GROUNDING

We have the option to run an extra electrode from the array. Visualize running an extra grounding electrode conductor to a ground rod from your array down the side of your building. In the 2014 NEC, it appeared this was required, most people said it was bad grounding science and did not do this. Now it is just an option.

690.51: MARKING OF MODULES

Modules must have the following marks on them (almost every module sold has these marks on the label):

- Open-circuit voltage Voc
- Operating voltage Vmp

• Maximum permissible system voltage	Maximum system voltage
• Operating current	Imp
• Short-circuit current	Isc
• Maximum power Power of module	(Vmp × Imp)
• Polarity	+ or –
• Max. series fuse rating	Max. fuse size

Figure 3.3 PV module label photo, courtesy of Sean White

www.axitecsolar.com

Model
AC-250P/156-60S

AXITEC
high quality german solar company

Pmpp: 250 Wp (+4.99Wp)
Umpp: 31.45 V
Impp: 7.98 A

Uoc: 37.90 V
Isc: 8.65 A

Max. System Voltage:	DC 1000V
Max. Series Fuse Rating	15A
Flame Rating	Class C

Intertek
4005942

Field wiring:
cooper only 12AWG min.
insulated for 90° C min.

Conforms to UL STD No.1703
Certified to ULC/ORD Std. C1703

All technical data at Standard Test Condition (STC)
Irradiance Level 1000W/m², Cell Temperature 25°C, Spectrum AM 1.5

Warning electric hazard
This module exposed to sunlight generates high voltage and current. Follow all safety precautions. Before installation, operation and maintenance, be sure to read and understand the instruction manual.

Précaution: Danger électrique
Ce module est exposé au soleil génére la haute tension et de courant. Suivez toutes les consignes de sécurité. Avant installation, exploitation et maintenance, veillez à lire et comprendre le manuel d'instruction.

AXITEC, LLC
160 Greentree Drive, Suite 101
Dover, Kent County
Delaware 19904
United States of America
info@axitecsolar.com

PNo:088/b
V140404

All modules used in the USA and Canada should be listed to the "UL 1703 Standard" by a Nationally Recognized Testing Lab (NRTL). NRTLs can be found through the OSHA website. (Internet search OSHA NRTL.) Inverters, combiner boxes, and charge controllers have to be listed to UL 1741. Racking can be listed to UL 2703. UL is a non-profit organization that makes up the tests, but is not the only NRTL that can perform the test. Other organizations that commonly perform the UL tests are CSA, ETL, Intertech, TUV, etc. Many large module manufacturers have in-house UL and other labs inside the factories. The best test is shooting a 1" round ice ball at the center of the module at 52 miles per hour. The modules are very strong!

690.52: MARKING OF AC MODULES

AC PV modules will not be required to have dc specifications on them. An ac module is essentially a dc module and a microinverter in one. It is only officially an ac module if it was listed as an ac module.

The following markings must be on ac modules:

- ac voltage
- frequency
- maximum power
- maximum current
- maximum OCPD
- identification of terminals or leads.

690.53: DIRECT-CURRENT PHOTOVOLTAIC POWER SOURCE

A permanent label at the dc PV power source dc disconnect indicating:

1. Max. system voltage from 690.7.
2. Max. circuit current from 690.8.
3. Max. rated output current of charge controller (if applicable).

If there is more than one power source, then each power source shall be marked (such as multiple MPPs on an inverter).

Another **dc disconnect** marking requirement is for a **plaque from 690.17** that says that the **line and load sides of the disconnect may be energized in the open position.**

Figure 3.4 Typical DC disconnect PV circuit specifications label, courtesy of pvlabels.com

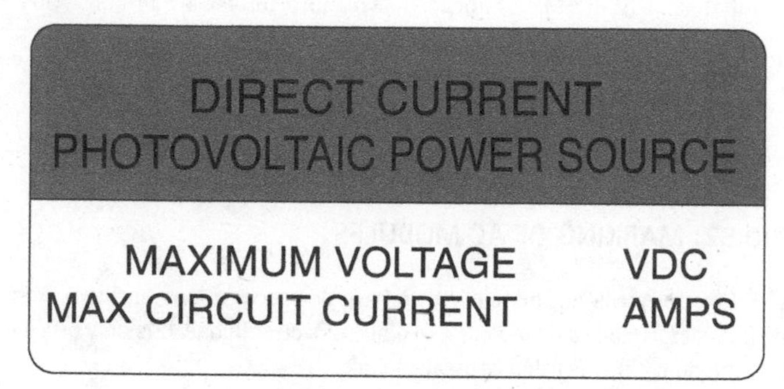

690.54: INTERACTIVE SYSTEM POINT OF INTERCONNECTION

Interactive system point of connection must be marked at disconnecting means indicating rated ac voltage and current.

Figure 3.5 Interactive System Interconnection Label, courtesy of pvlabels.com

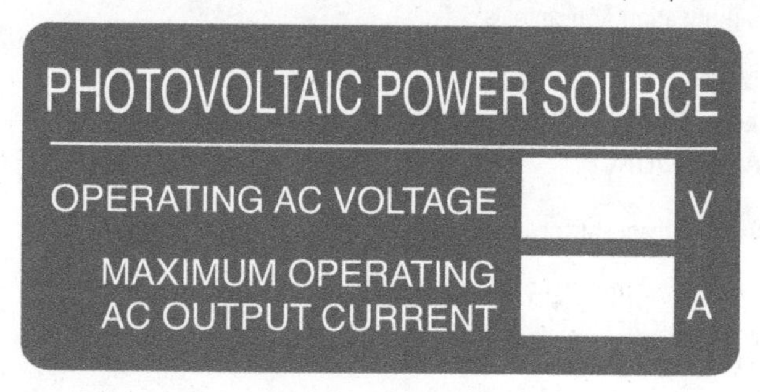

690.55: PV SYSTEMS CONNECTED TO ENERGY STORAGE

PV system output conductors shall be marked to indicate polarity (positive and negative).

690.56: IDENTIFICATION OF POWER SOURCES

690.56(A): FACILITIES WITH STANDALONE POWER SYSTEMS

Off-grid systems must have a **sign** outside the building in a visible place **indicating where the disconnect is located.**

690.56(C): BUILDINGS WITH RAPID SHUTDOWN

Rapid shutdown was explained earlier in this chapter in 690.12 on page 54.

> It can be confusing that the sign required for systems that comply with 690.12 Rapid shutdown is not specified in 690.12 Rapid shutdown. Instead it is specified later in the code. Determining where to look when unprepared can cause NEC paper cuts. Be prepared!

Let us show you the label first, which is very self-explanatory and then discuss the label.

Figure 3.6 960.56(C)(1)(a) Label for shutting down conductors inside the array, (module level shutdown)

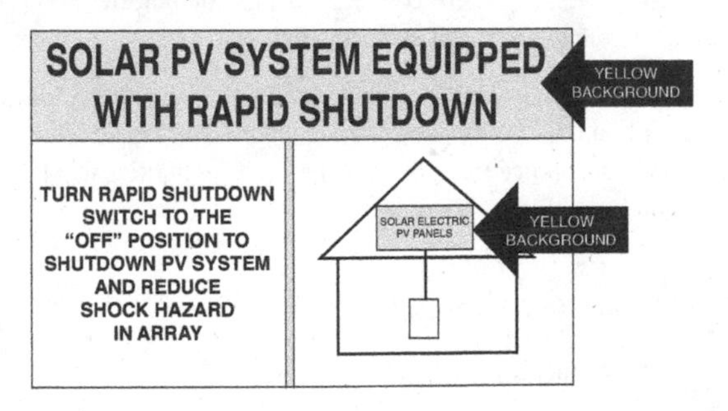

Figure 3.7 960.56(C)(1)(b) Label for shutting down conductors leaving the array (not inside the array)

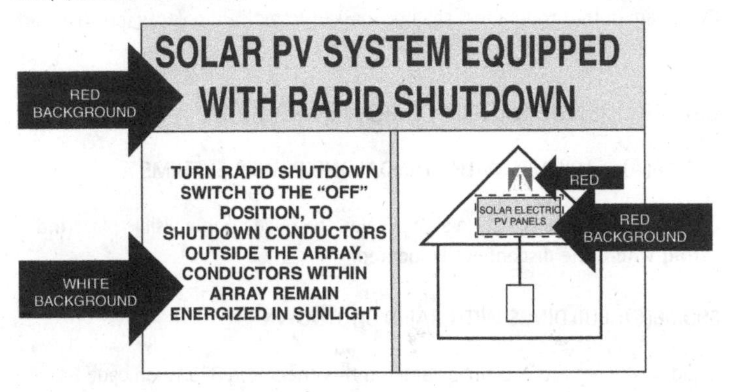

This book and the NEC are in black and white; however, there are coloring requirements for this sign.

The sign that shuts down conductors inside the array is safer and has yellow coloring.

The sign that does not shutdown conductors inside the array is more dangerous and hence colored red with dotted lines around the array where voltage can be higher. This sign also includes an exclamation mark.

The rapid shutdown shall be no more than 3 feet from the service disconnecting means where the PV systems are connected and shall indicate the location of rapid shutdown switches if not at same location.

The rapid shutdown sign will let Fire Department first responders know that it is safer to walk into a burning building than if there was no rapid shutdown and this will keep our insurance rates from going up if firefighters feel it is safe to put out fires on houses with PV on them.

690.56(C)(3): rapid shutdown switch

Switch shall have a label within 3 feet that says:

Figure 3.8 Rapid shutdown switch, courtesy of pvlabels.com

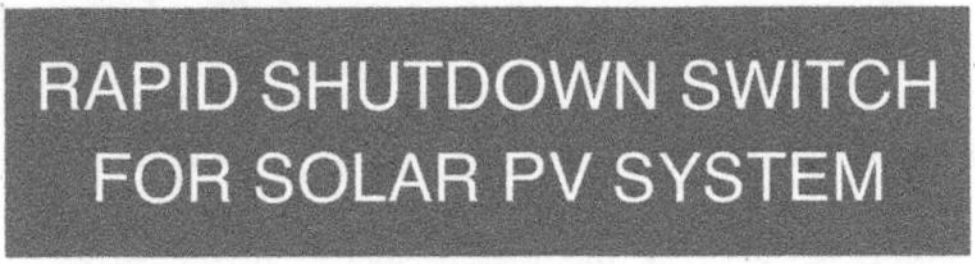

RAPID SHUTDOWN SWITCH
FOR SOLAR PV SYSTEM

Part VII/690.59: connection to other sources

PV systems connected to other sources shall comply with Article 705, which we will cover in Chapter 4 of this book on page 71.

PART VIII: ENERGY STORAGE SYSTEMS
690.71: GENERAL

Energy storage systems connected to a PV system shall comply with Article 706 Energy Storage Systems.

Article 706 Energy Storage systems is new in the 2017 NEC and in previous editions of the NEC we were using Article 480 Storage Batteries. It has been said that NABCEP is still using questions related to Article 480 on 2017 NEC exams, so it is good to be familiar enough to look things up in Article 480 as well as Article 706. I agree, it is not straightforward.

Also note that in the 2014 NEC and earlier, that energy storage systems were often part of a PV system. Now they are considered separate systems with different rules, such as rapid shutdown is not required for energy storage systems.

690.72: SELF-REGULATED PV CHARGE CONTROL

We can avoid using a charge controller if we size the PV and battery, so that there will be no overcharging. The rules are:

1. PV source circuit voltage and current is matched to the battery.
2. Maximum charging current in 1 hour is less than 3% of the battery capacity (this way we will not overcharge).

Article 705: interconnections

ARTICLE 705: INTERCONNECTED ELECTRIC POWER PRODUCTION SOURCES

The main part of Article 705 that we will study will be Section 705.12: point of connection, which is where all of the complicated bidirectional literature lies. It is different from the rest of the code, because we have power flowing in both directions; loads, sources, and anti-loads! Which is just like what utilities do with multiple power plants, but on a different scale.

Anti-loads can be convenient when understood. They cancel out loads and can decrease current on a line (which is why they decrease your bill). Anti-loads can also decrease voltage drop, and, if installed correctly, can make unbalanced three-phase power more balanced. Do not expect everyone to understand this last statement. People are so used to unidirectional power.

As more intermittent renewable energy is connected to the grid and as more electric vehicles need to be charged as they arrive home from work, there will be different demands put on the grid that need to be stabilized. In the future we will use renewable energy to stabilize the grid by being able to affect the power quality and by being able to store energy and send it back in packets that the grid needs to be stable.

705.10: DIRECTORY

A sign should be at each service equipment location and at each power production source showing where every other service and power production source is.

705.12: POINTS OF CONNECTION – SUMMARY

705.12(A): supply-side connection

This is between the main breaker and the meter.

The sum of the overcurrent devices backfeeding the supply side shall not exceed the rating of the service. It would be rare to want to have more interconnected PV sources than the rating of the service. That would be a lot of PV!

705.12(B): load-side connections

705.12(B) is called **utility-interactive inverters**, but it is referring to load-side connections.

Types of load side connections are:

1. feeders
2. taps (tapping a feeder)
3. busbars (120% rule goes here).

Usually, we are talking about supply-side and load-side connections.

705.12: POINTS OF CONNECTION – DETAILS

705.12(A): supply-side connections

You can connect as much as the service can handle.

On a 705.12(A) supply side connection, you can add as much PV as the service can handle, which is much more than you would probably ever want to connect in a net-metering situation.

The reason we do not consider the loads or the main breaker (main service disconnect) here is because the main breaker is protecting the loads from the utility. Since the utility can have a lot of fault current and the main breaker is designed to protect the main service panel from the utility, then by adding solar we do not have to be concerned about the extra current from the solar on the other side of the main breaker.

That would mean that on a 200A service at a house, we could install at least a 200A PV breaker.

If someone were trying to install a 200A solar breaker on a house with a 200A service, they would be producing much more than they would ever be using. For most purposes, you can add as much solar as you will ever want to install using a supply-side connection.

The reason we do not always install on the supply-side is that it is much easier to pop in a breaker on the load-side. On the supply-side, there is no easy way to turn off the power and we often have to pull the meter or coordinate with the utility to make the connection.

We never want to work on anything while it is hot (energized) when working with PV!

705.12(B): utility-interactive load-side connections

705.12(B) is the complicated (exciting/not boring) part of Article 705.

705.12(B)(2): This is where the fun happens.

125% of the inverter output current is what we use for most 705.12(D) load-side calculations.

In the past (2011 NEC and earlier) we would use the inverter OCPD rating for our calculations. Using 125% of the inverter output current can give us a little more leeway than in the past, since when sizing a breaker for an inverter output circuit, we round-up to the next common OCPD size (if under 800A).

705.12(B)(2)(1): feeders

Easy typical feeder definition: The conductors that feed the subpanel or branch circuit OCPD. Example: a feeder going from the main service panel to a subpanel.

Feeder NEC definition: All circuit conductors between the service equipment, the source of a separately derived system, or other power source, and the final branch-circuit overcurrent device.

Article 100: definitions is where you can find many definitions and is near the front of the NEC.

If we are connecting two power sources to a feeder, there will be a place on the feeder subject to the currents of both sources added together. We must make sure that we protect the feeder. There are two ways that we can protect the feeder:

1. Make the portion of the feeder that is subject to the extra current larger.
2. Install an OCPD on the feeder that will protect it.

The safest place to put the OCPD on the feeder in Figure 4.3 is where the inverter connects to the feeder at the source of the overcurrents, as shown in Figure 4.3. It is generally accepted that you can place the overcurrent protection within 25 feet of the inverter connection point to the feeder, because of the 25 feet tap rule, which we will learn about next.

705.12(B)(2)(2): taps

WARNING: This may be the most complicated part of this book. Top experts argue about PV feeder taps. It is suggested that you do not let yourself get stuck here. Scan through this and come back to it after you understand everything else. If you understand PV feeder taps, you will be part of an elite club and you will be extremely valuable. This material has never been on a NABCEP exam ... yet – to the best of my knowledge.

Figure 4.1 705.12(B)(2)(1), two sources on one feeder

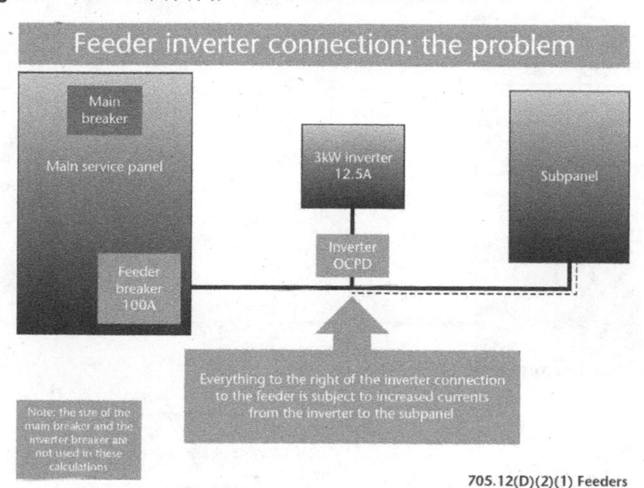

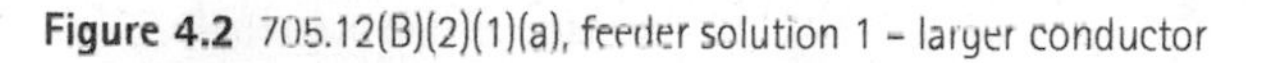

Figure 4.2 705.12(B)(2)(1)(a), feeder solution 1 – larger conductor

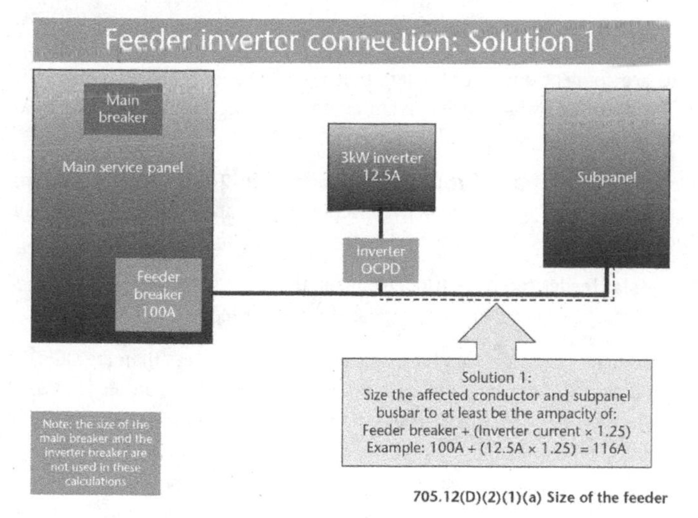

Figure 4.3 705.12(B)(2)(1)(b), feeder solution 2 – overcurrent protection device

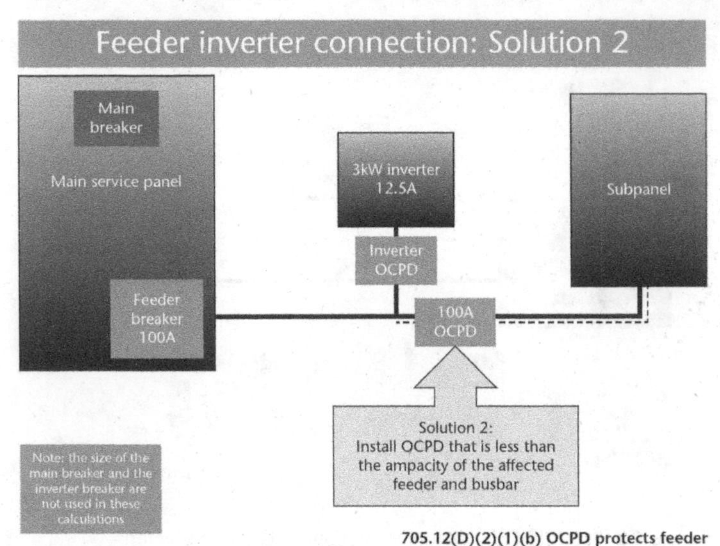

Taps are connections onto feeders that are not completely protected with OCPD. When we include solar in a tap, according to Section 705.12(D)(2) (2), we have to follow the "tap rules."

We will learn next about regular non-solar taps in **240.21(B): feeder taps**. Before we go into utility-interactive inverter feeder taps, let us get warmed up by digesting regular non-solar feeder taps:

Non-solar feeder tap rules. (Understanding this is the first step to understanding the more complicated utility-interactive inverter feeder tap rules.)

The danger: Feeder taps can have ampacity that is a lot less than the OCPD at the source of the feeder from which they are tapped. This can be somewhat analogous to a service, where there is no overcurrent protection for the service entrance conductors entering your meter.

240.21(B)(1): the 10 feet tap rule (for taps of 10 feet or under)

- Tap ampacity > tap load
    - Ampacity of tap conductor has to be greater than the load.
- Tap ampacity > equipment containing tap OCPD
    - That is the electrical box containing the tap conductors.
- Tap conductors in conduit

Figure 4.4 240.21(B)(1), non-solar ten-foot tap rule. The tap conductor must be at least 10% of the feeder breaker

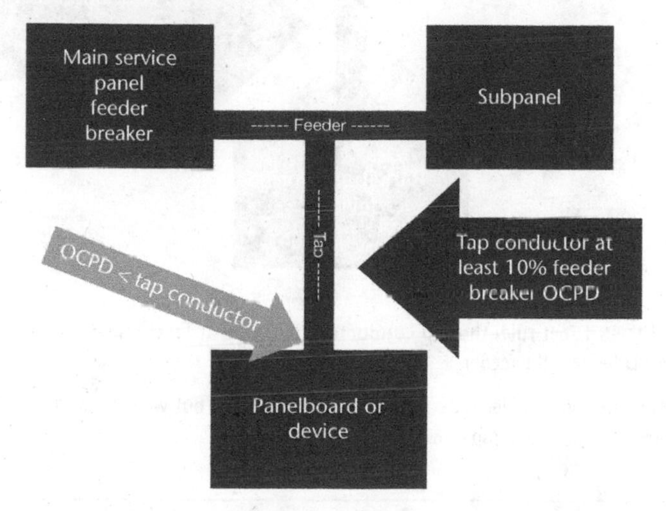

For the **10 feet rule, the tap conductor has to be at least 10% of the OCPD** feeding the feeder.

240.21(B)(2): the 25 feet tap rule (for taps of 10–25 feet)

- Tap conductors terminate on a single OCPD.
- They are protected by a raceway (conduit).

Figure 4.5 240.21(B)(2), non-solar 25-foot tap rule. The tap conductor must be at least one-third of the feeder breaker

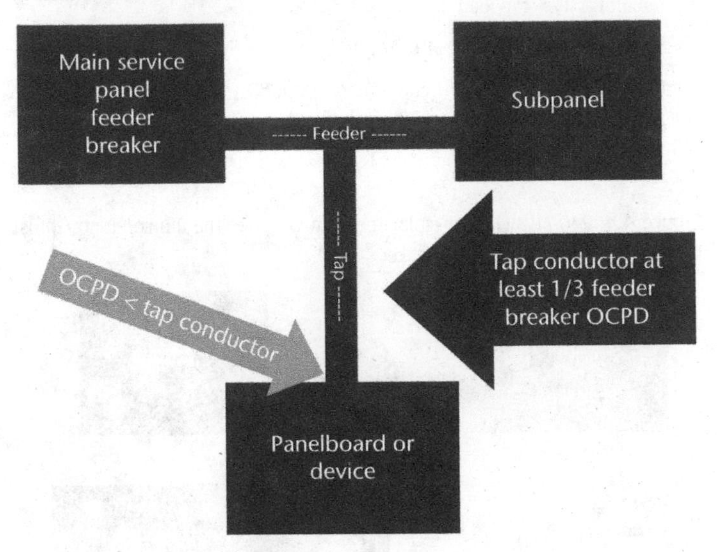

For the **25 feet rule, the tap conductor has to be at least one-third of the OCPD** feeding the feeder.

There are more feeder tap rules in Section 240.21(B), but we will focus on the most popular of the tap rules.

> Why would the NEC allow us to have tap conductors that are not protected from the source with OCPD? Most experts do not have an answer. Just think of your meter. If you stuck a screwdriver between the conductors at the meter, you would likely end up with a melted screwdriver or worse.
>
> We protect tap conductors in a raceway and there are limitations on the length of the conductors. A longer conductor has more resistance.

Other examples of conductors that are not protected from direct shorts are PV output circuits. If we short circuited the dc inverter input conductors at a MW inverter, we would not have OCPD that would open the circuit. That is because according to the NEC, those OCPD devices would have to be at least 156% of the short circuit current. That is because PV Isc is not much more than PV Imp.

In the past, solar installers called the "supply-side connection" a "line-side tap," which was incorrect, since with a supply-side connection we do not follow the tap rules. Supply-side connections are more like adding a new service.

705.12(B)(2)(2) and 240.21(B): feeder taps (aka "the tap rules for solar")

Summary: With the solar tap rules, add 125% of inverter current to the feeder source breaker for feeder tap calculations.

With solar feeder taps, we need a larger conductor than with typical feeder taps.

In the image in Figure 4.6, you can see that there are a lot of things going on and more than one thing we have to check for, however, most of the time, it will be obvious which rule is the limiting factor.

Figure 4.6 Codes on a feeder

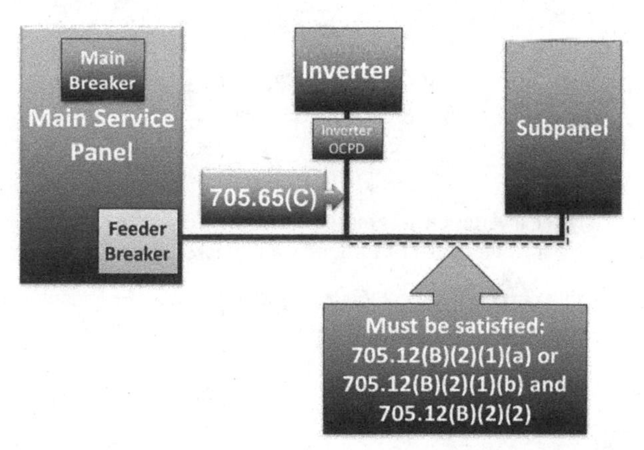

When applying the feeder tap rules to solar, after connecting the inverter to the feeder, we are applying the tap rules of 705.12(B)(2)(2) to the part of the feeder going to the loads, not the inverter output circuit. The inverter output circuit can only have the currents of the inverter.

Sizing the affected conductor:

- Add the feeder breaker amps to 125% of the inverter amps.
- Multiply by:
  - 10% or 0.1 if the feeder is under 10 feet
  - 33.3% or 0.333 if the feeder is 10–25 feet.
- The conductor may not be smaller, but may be larger than the calculated sizes.
- The other rules for non-solar feeder taps also apply.
- Rules for solar feeders that are not taps also apply.

Example

- 10 feet tap rule.
- 200A source breaker.
- 16A inverter.

The conductor must be at least 10% of (source breaker + 125% inverter current):

16A inverter × 1.25 = 20A
20A + 200A source breaker = 220A
0.1 × 220A = 22A
Conductor must be at least 22A

Additionally, the inverter breaker must be less than 22A and for a 16A inverter; we would use a 20A breaker (16A × 1.25 = 20A breaker).

One more thing to check is the 230A feeder, which needs to be enough to carry currents from the source and 125% of the inverter.

To summarize solar and feeder taps

If there are two sources that can supply currents to a feeder, we have to add those sources together when determining the minimum size of the feeder tap conductor. When we add the two sources together, we add the breaker on the utility supply-side of the feeder to 125% of the inverter current.

We then multiply by 10% if the portion of the feeder is under 10 feet to get our minimum conductor size.

If the portion of the feeder is between 10 and 25 feet, we multiply by one-third to get our minimum conductor size.

705.12(B)(2)(3): busbars

There is a lot of information in **705.12(B)(2)(3)**, so pay close attention.

705.12(B)(2)(3)(A): NOT EXCEEDING THE BUSBAR AMPACITY

You can put the solar breaker(s) anywhere on the busbar as long as 125% of the inverter current plus the current of the breaker protecting the busbar (main breaker) does not exceed the ampacity of the busbar.

You do not have to put the solar breaker at the opposite side of the busbar from the supply breaker if you do not exceed the rating of the busbar as we do with the 120% rule, which we will discuss next.

This also applies to subpanels.

You can put the breakers anywhere on the busbar if you do not exceed the rating of the busbar.

705.12(B)(2)(3)(B): THE 120% RULE (EXCEEDING THE BUSBAR AMPACITY)

Reminder: In the 2011 NEC and earlier we used the inverter breaker ampacity in the 120% rule calculations. Now we use the ampacity of the **inverter × 1.25**.

Figure 4.7 Main + solar less than the busbar rule – some people call this the 100% rule

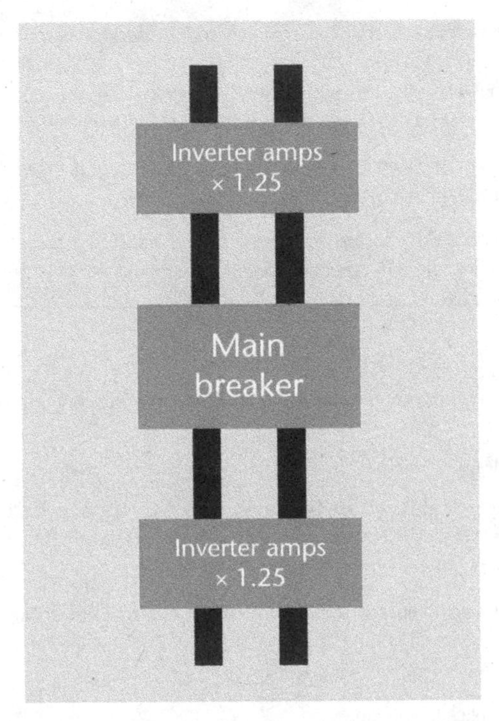

The main breaker plus 125% of the inverter current cannot exceed 120% of the rating of the busbar. When exceeding the busbar ampacity, the solar breaker shall be on the opposite side of the busbar from the main breaker.

> Usually the main breaker is at the top of the busbar. If this is the case, then the backfeed breakers would have to be at the bottom (opposite) side of the busbar. Center-fed busbars can use the 120% rule as long as the solar backfeed breaker(s) are located on one far end of the busbar.

Often we walk up to a house and want to see how much inverter we can install. With this method, we take note of the main breaker size and the busbar rating.

Figure 4.8 The 120% rule

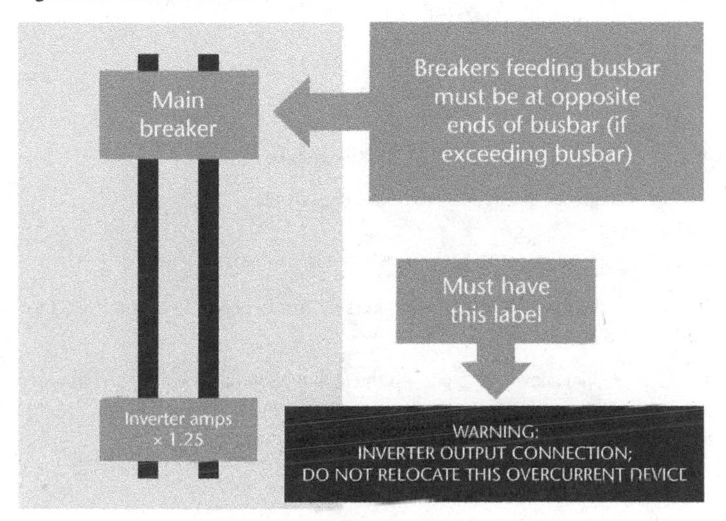

$$(\text{Busbar} \times 1.2) \geq \text{Main} + (\text{Solar} \times 1.25$$
$$(\text{Busbar} \times 1.2) - \text{Main} \geq (\text{Solar} \times 1.25)$$
$$((\text{Busbar} \times 1.2) - \text{Main})/1.25 \geq \text{Solar}$$

Then you have your max inverter size in amps.

Calculator shortcut for the 120% rule.

On your calculator enter:

Bus × 1.2 =
– Main =
/ 1.25 =

Then you have your max inverter size current.

Example

Busbar = 200A
Main = 200A

What is your largest inverter size?

$$200A \times 1.2 = 240A$$
$$240A - 200A = 40A$$
$$40A/1.25 = 32A$$

Largest inverter is 32A

You can multiply by the ac voltage to get the power!

705.12(B)(2)(3)(C): SOLAR + LOAD BREAKER METHOD (AKA "THE SUM RULE")

The sum of the load and supply breakers do not exceed the busbar (this also works for subpanels).

705.12(B)(2)(3)(c) allows us to add up the load and supply breakers and as long as they don't exceed the rating of the main supply breaker (or the busbar rating) then we are good.

$$\text{Solar} + \text{Load} \leq \text{Busbar}$$

Note that for the sum of the **solar + load breaker** accounting method, **we do not use 125% of the inverter current**, we use the size of the breaker. This is easy and convenient when combining inverters with an ac combiner subpanel. The intention of the sum rule was to make it easier to do ac combining, without the restrictions of the 120% rule.

705.12(B)(3): marking (point of connection)

If the busbar is supplied by multiple sources, then equipment must be marked to indicate the sources.

If we have a panelboard that is fed by multiple sources, we need to let people know that power is coming from more than one direction.

705.12(B)(4): breakers must be suitable for backfeed

It is assumed that fused disconnects are suitable for backfeed.

Figure 4.9 Sum of the solar breaker + load breaker rule

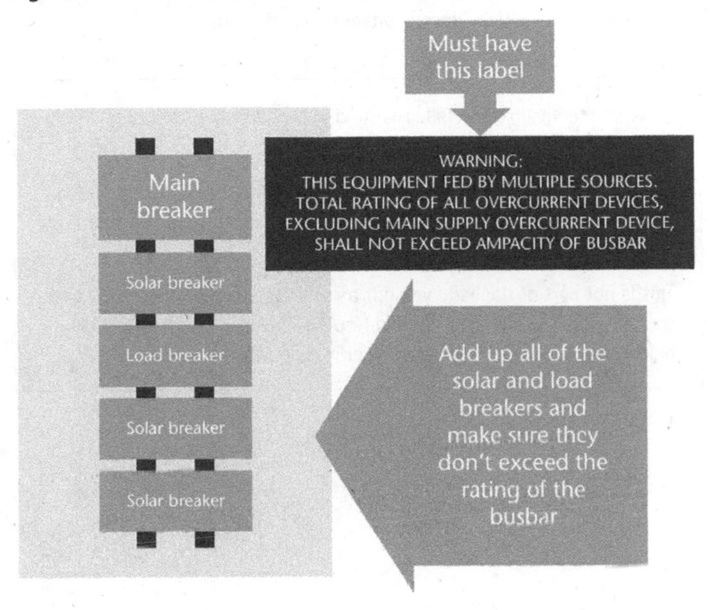

GFCI and AFCI breakers are suspect and breakers marked line and load are not suitable for backfeed.

705.12(B)(5): fastening

Breakers used for utility interactive applications do not need fasteners.

Fasteners are used when a breaker could pop loose and you would have an energized breaker flopping around. Since grid-tied inverters have anti-islanding provisions and immediately shut down when there is an outage, we do not have to be concerned about an inverter feeding power to a breaker that comes loose.

705.100: UNBALANCED INTERCONNECTIONS

Single-phase inverters are to be connected to three-phase systems to limit unbalanced voltages to less than 3%.

Three-phase inverters need to turn off with loss of voltage in a phase unless the system is designed so unbalanced voltages will not result.

> Solar systems can be installed with single-phase inverters onto three-phase systems that will actually fix unbalanced voltages.
>
> The technology is there for three-phase inverters to counteract unbalanced voltages by pushing unbalanced current onto phases with low voltage.
>
> This is not part of the code yet, but as solar gets bigger and utilities see the potential for inverters to condition power, solar inverters will play a bigger part in cleaning up dirty power in more ways than one.

Chapter 5

Safety, strategy and investing your time wisely

There is enough safety material that could be covered on the NABCEP Certification Exam that you could spend the rest of your life studying for it. The key here is to spend your time making wise choices.

Typically with safety, you want to err on the side of being the safest. If you are given a choice between inspecting your multimeter, using your multimeter, and touching wires, then you should inspect your multimeter before you use it.

If you were given a choice between turning a system off and checking voltage when something is turned on, it would be wise to check something when it is turned off first before checking it when it is operating. You should conduct the least invasive procedures first.

Before you qualify to take a NABCEP Certification Exam, you are required to have 10 hours of OSHA training. It is good strategy to take the OSHA training closer to your exam date, rather that leaving a lot of time between your OSHA training and your exam, so it will be fresher in your memory. Some students take 30 hours of OSHA training, when only 10 hours is (officially) required.

Here are a few safety pointers that are common on many construction exams:

1. Use fall protection when you can fall 6 feet or more.
2. PFAS is a personal fall arrest system, which consists of an anchor, a lanyard, and a harness.
3. An extension ladder should have a 1:4 ratio of distance to height.
4. Never use a painted wooden ladder (paint can hide defects).

5. Much of the responsibility of safety falls on the employers, but not everything. Employers have to teach about safety hazards.
6. Employers provide most of the safety equipment.
7. The person who puts the lockout tagout (LOTO) in place should be the person to remove it.
8. Extreme dangers are from arc flashes.
9. The safest way to measure current is with a clamp ammeter.
10. A safety monitoring system is when a designated safety person watches another employee and tells them if they are getting close to the edge. The designated safety person must be on the same level as the employee, close enough to communicate orally, and must not perform other duties when watching.
11. Add acid to water so acid doesn't splash.
12. For lightning, use a lightning protection system and surge protection.

Many people studying for the NABCEP Exam are memorizing OSHA subparts. You are better off learning more practical safety procedures. There is only so much safety you can learn in time for the exam. Use common sense. It does not make sense that NABCEP would test you on memorizing that OSHA CFR 1926 Subpart M is for fall protection; however, it would be a good idea to read subpart M a month before the exam, since fall protection is so important. All OSHA materials are free online.

Other areas of common sense are covering subjects such as AHJ questions, managing the project, commissioning, etc. For most of these types of question, you have to rely on your knowledge of the NEC, solar, and common sense. You need to be a good decision maker to pass this exam and not just a good memorizer. Make sure to go into the exam prepared, well rested, calm, and ready to make good decisions.

The goal of test writers should not be to trick people into failing. They just want you to demonstrate that you would be someone that the good people at NABCEP will be proud to call "Certified"!

Wire sizing

Chapter 6

PV source circuit wire sizing – abbreviated

1. Continuous current (Isc × 1.56) and check terminal/wire temperature.
2. Maximum current (Isc × 1.25) and conditions of use.
3. Is wire large enough for OCPD in 240.4?

PV source circuit wire sizing – more detailed

1. Continuous current and check terminal/wire temperature
   a. Check in table 310.15(B)16 or 310.15(B)17 in the column of the temperature of the wire or the terminals, whichever is less.
2. Maximum current and conditions of use
   a. Derate for conditions that cause heat, which are:
      i. 310.15(B)**(2)(a)**: hot ambient temperature table
      ii. 310.15(B)**(3)(a)**: more than three wires in conduit table
      iii. **Note**: 310.15(B)(3)(c): distance conduit over roof table, was taken out of the NEC in 2017
   b. Make sure this corrected ampacity does not exceed OCPD, and, if it does, chose next size larger conductor.
3. Is wire large enough for OCPD?
   a. Take derated wires and terminals and round-up to next OCPD size. Compare theoretical OCPD size to real OCPD and if real OCPD is larger than theoretical OCPD, use larger wire.

Other circuits are sized the same way, just use appropriate maximum current and 125% greater continuous current.

Why PV source and output circuits are different from normal wire sizing:

1. We size from short circuit current and not operating current.
2. We multiply by 1.25 an extra time because of irradiance beyond STC.

PV is unusual because it creates power based on environmental conditions that can be greater than testing conditions.

Note: Wire sizing may be done differently under engineering supervision with a licensed electrical engineer.

Typical wire sizing for PV systems (excluding PV source and output circuits)

1. Operating current × 1.25 and check terminal temperature
2. Operating current and conditions of use/check OCPD
3. Is wire large enough for OCPD?

WIRE SIZING COMPLEXITY

Before you get a headache, be aware that wire sizing is complicated, and, if you are just trying to pass a NABCEP Certification Exam, there is a good chance you will see minimal wire sizing questions, such as one easy one. This may be due to the fact that the exam committees cannot agree how to properly size a wire. The NEC is complex and interpreted by many opinions. Do not feel bad if you find wire sizing difficult. If you do find it difficult, join the club.

In this chapter, we will go over the different calculations and theories for sizing conductors using the tables in the NEC.

EXAMPLE AC INVERTER OUTPUT CONDUCTOR SIZING

First, we determine the current of the inverter. The easiest way to calculate the output current is using the formula:

$$W = VI$$
$$I = W/V$$

For example, if we have a 3.8kW inverter on a house then:

$$I = 3800W/240 = 15.8A \text{ inverter}$$

It is best to get the inverter current from the inverter datasheet. If you are not given these data, this is how to calculate the current. Sometimes the current from the datasheet can be slightly different from the calculated current.

To size the OCPD (breaker) for the inverter, we would multiply by 1.25:

$$1.25 \times 15.8A = 19.8A$$

125% of output current is called continuous current.

PV is always considered continuous, since it goes for 3 hours or more.

They do not make 19.8A breakers and the rule is to round-up to the next common breaker size, which is a **20A breaker**.

We can then go to **NEC 240.4** and see that a **20A breaker requires at least 12AWG** copper wire. If we used anything smaller, the 20A breaker would not protect it and it could be a fire hazard.

Terminals are the connection from the wire to the breaker or device. If the wire heats up, it will heat up the terminal and the wire acts as a heat sink taking heat away from the terminal.

Then, we take our continuous currents and check for **terminal temperature limits**. Terminals are typically rated for 60C and 75C. Under 100A we assume 60C rated terminals unless it is marked otherwise. Rooftop equipment is often rated above 60C.

Assuming that the ac inverter output conductor is running in conduit, we will use Table 310.15(B)(16).

Table 6.1 From Table 310.15(B)(16) Allowable ampacity for conductors in conduit at 30C

Size of wire	75C rated conductor	90C rated conductor
14AWG	20A	25A
12AWG	25A	30A
10AWG	35A	40A

If our wire were 90C rated THWN-2 and connected to **75C terminals**, we would still have to check the **75C column** when checking for terminals.

In this case, we could have used a 14AWG wire, since a **14AWG conductor can carry 20A** and we only need it to carry 19.8A, but as we determined above, Section 240.4 required that a 20A breaker uses at least 12AWG wire, so we need at least 12AWG.

Next we are going to check for conditions of use, which are temperatures and conditions that cause the conductors to heat up.

If the inverter were on a rooftop, with two different inverter circuits running in the same conduit, in a location with a **high design temperature of 40C**, then we will have to account for these conditions.

Then we go to Table 690.31(A), which has the same information for our purposes as Table 310.15(B)(2)(a).

This time we are using the **90C column** for the conductor and in this case **40C** in the 90C column gives us a **derating factor of 0.91**.

We are going to find one more condition of derating for greater than three current-carrying conductors in conduit, since we have two different inverter circuits running in the conduit across the roof. Let's assume they are both 3.8kW inverters.

Two inverters will give us four current-carrying conductors (we do not count the neutrals of 120/240V ac inverters, since they carry no current).

Table 6.2 From Table 690.31(A) or 310.15(B)(2)(a) Ambient temperature correction factors

Ambient temperature	75C conductor	90C conductor
26–30C	1	1
31–35C	0.94	0.96
36–40C	0.88	0.91
41–45C	0.82	0.87
46–50C	0.75	0.82
51–55C	0.67	0.76

Table 6.3 From Table 310.15(B)(3)(a) Derating for more than three current-carrying conductors in a raceway or cable

Number of conductors	Derating factor
4–6	80
7–9	70

In this case, the **derating is 80%.**

We now have two different derating factors for conditions of use, the **0.91** and the 80%. It is easy to turn the 80% into a decimal by dividing by 100 and we have a derating factor of **0.80**.

Our combined derating factor is calculated by:

$$0.80 \times 0.91 = 0.728$$

Then we go back to the current of the inverter, which was 15.8A.

Since it is hot in that conduit, the ampacity of the conductor needs to be greater; to make 15.8A larger we will divide by the derating factor of 0.728:

$$15.8A / 0.728 = 20.6A$$

So we need a conductor of at least 20.6A in this case. We will look back to Table 310.15(B)(16).

Table 6.4 From Table 310.15(B)(16) Allowable ampacity for conductors in conduit at 30C

Size of wire	75C rated conductor	90C rated conductor
16AWG	---	18A
14AWG	20A	25A
12AWG	25A	30A
10AWG	35A	40A

In this check, in the 90C column a 14AWG wire can carry 25A, so we could use a 14AWG conductor here, although as before, the 75C terminals with a 20A breaker require at least a 12AWG conductor, so we still have to use at least a 12AWG wire because of the previous check.

We can then double-check the 12AWG wire at conditions of use to make sure it is protected by a lower ampacity OCPD 12AWG wire at conditions of use is 20.6A (as above).

20.6A wire is allowed to be protected by next size larger OCPD, which is 25A.

(Even though the wire is derated to 20.6A, we can round-up to the next size OCPD of 25A, since the wire is as good as the next greater size OCPD according to the code, which is confusing, because it is not logical).

25A > 20A OCPD we are using, so this checks out and 12AWG works!

SUMMARY

1. Determine the current and multiply by 1.25 for continuous current.
2. Round-up to common breaker size.
3. Make sure the breaker can protect the conductor (Section 240.4 gives the wire size).
4. Check for terminal temperature ampacity with continuous current in the terminal temperature column of the table – get wire size.
5. Do conditions-of-use deratings and use non-continuous current in wire temperature column – this gives wire size.

6. Take the conditions of use corrected ampacity of the conductor and round-up to the next common OCPD. If this OCPD size (which is really the ampacity that the conductor can protect) is larger than the OCPD, then we are good to go. If this OCPD size is smaller, then size up your conductor and try again. (The OCPD size here is not an OCPD, but is the theoretical rounded-up ampacity, since we can round-up, we are taking advantage of the ability to round.)
7. Pick the largest conductor.

There are more OCPD checks in 240.4, but the method here almost always satisfies.

If we were going to size wires for PV source and output circuits, we would use Isc × 1.25 for current and Isc × 1.56 for continuous current. With PV source and output circuits, everything uses 1.56 except for the conditions-of-use calculations.

Following is an example PV source circuit wire sizing calculation:

- Isc = 8.5A
- Current carrying conductors in conduit = 8
- Hot design temperature = 35C
- Combiner box terminal rating = 75C
- Fuse in combiner = 15A
- THWN-2 conductors in conduit

1. Small conductor rule from Section 240.4:
    - 15A fuse requires at least 14AWG
2. Terminal temperature limits and continuous current:
    - Isc × 1.56 = 8.5 × 1.56 = 13.3A

    According to Table 310.15(B)(16), the smallest 75C conductor that we can use is **14AWG**.
3. Conditions-of-use calculations:
    - 35C design temperature.
    - From Table 690.31(A) or Table 310.15(B)(2)(a), derating factor is 0.96 (we do not use terminal temperature column with conditions of use, we use the 90C column, because THWN-2 is 90C rated).

- Since there are eight current-carrying conductors in conduit, we use Table 310.15(B)(3)(a): derating factor is 70% or 0.7.
- Multiply derating factors together: 0.7 × 0.96 = 0.672.
- Isc × 1.25 = 8.5A × 1.25 = 10.6A.
- Use derating factor to increase 10.6A by dividing: 10.6A / 0.672 = 15.8A.
- 15.8A is greater than 15A OCPD, so we do not need larger wire here.
- In Table 310.15(B)(16) in the 90C column, the conductor that can handle more than 15.8A is **16AWG**.
- OCPD conditions-of-use wire check by taking 15.8A and rounding up to next larger OCPD in theory, which is 20A. Since 20A is greater than 15A, then this check checks out (and is not logical).
- OCPD terminal check by taking terminal ampacity of 20A wire and rounding-up to overcurrent device size, which is 20A and since 20A is greater than OCPD size, then OCPD checks out here too.

Table 6.5 From Table 310.15(B)(16) Allowable ampacity for conductors in conduit at 30C

Size of wire	75C rated conductor	90C rated conductor
18AWG	---	14A
16AWG	---	18A
14AWG	20A	25A
12AWG	25A	30A
10AWG	35A	40A

Table 6.6 From Table 690.31(A) or 310.15(B)(2)(a) Ambient temperature correction factors

Ambient temperature	75C conductor	90C conductor
26–30C	1	1
31–35C	0.94	0.96
36–40C	0.88	0.91
41–45C	0.82	0.87

Table 6.7 From Table 310.15(B)(3)(a) Derating for more than three current-carrying conductors in a raceway or cable

Number of conductors	Derating factor
4–6	80
7–9	70

Table 6.8 From Table 310.15(B)(16) Allowable ampacity for conductors in conduit at 30C

Size of wire	75C rated conductor	90C rated conductor
18AWG	---	14A
16AWG	---	18A
14AWG	20A	25A

From this exercise, we can see that the small conductor rule and the terminal check for continuous current led to a 14 AWG wire and the conditions-of-use check led to a 16 AWG wire, so we have to go with at least a 14 AWG wire here. In the field, nobody uses a 14AWG conductor; however, it would not be a code violation to use a 14AWG conductor. Some of the reasons we often use wires that are larger than required are efficiency, also known as voltage drop, not causing trouble if the inspector does not like your using such a small wire, and often residential installers just carry 10 AWG wire on the truck. By the way, the smallest equipment grounding conductor allowed is a 14 AWG conductor as long as it is protected in conduit.

There will be a variety of conductor sizing calculations in the practice exam questions later in the book. The best way to do wire sizing is by repetition. Once you practice enough, you can do much of it with a calculator and writing down just a few numbers.

When a wire is in two places at once

If you have a conductor that is in two places at once, say on a rooftop and inside a building, you have to consider the ampacity of the conductor as the lowest ampacity.

Section 301.15(A)(2) states that when more than one ampacity applies for a given circuit, the lower of the ampacities applies, with one exception.

Exception: If the lesser ampacity is 10 feet or 10% of the length, then you can use the higher ampacity.

Less than 2 feet exception for greater than three current-carrying conductors in raceway.

If more than three current carrying conductors are in a raceway, then you do not have to derate if the raceway is less than 24".

One last comment about wire sizing: Many of the top experts have different ways of interpreting how to do wire sizing according to the NEC. You may also see some experts take the "conditions-of-use" into consideration when sizing OCPD at the end of the wire sizing exercise. If the derated wire does not exceed the rating of the OCPD, they will use a larger wire. You may choose to size your wire this way.

For the NABCEP Exams, I would not expect many wire sizing questions like this, since there are differences of opinion and the Exam is not designed to pass people who pick the right philosophy.

It is my job to present the facts, the code, and to inform you that there are different ways of doing things and to let you decide for yourself, which way to apply the NEC, especially when the top experts disagree.

Solar trigonometry

Knowing trigonometry can help you score a few points on exams and in real life. Many people are overwhelmed by trigonometry. You can survive without a few points, but once you understand trigonometry, you will have better confidence.

Trigonometry is using the relationships of triangles and determining everything about the triangle from only knowing two different things about it.

If we know two of the four things about a right triangle, we can figure out the rest.

The four things are the length of the three sides of the triangle and an angle (besides the 90° angle of a right triangle).

Let's look at some triangles and relationships.

Figure 7.1 Right triangle relationships

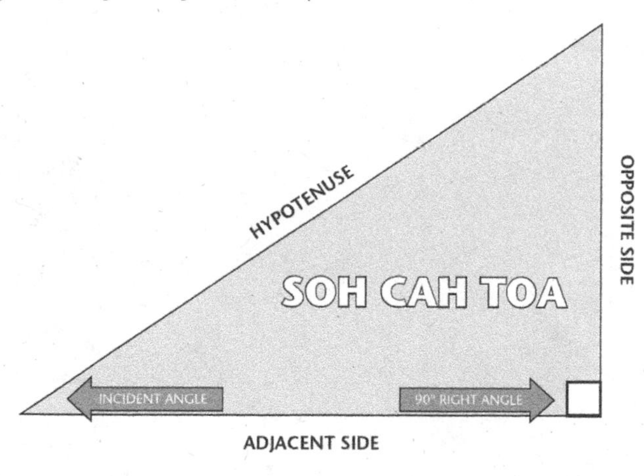

SOH CAH TOA is a mnemonic that is used for remembering the relationships of the triangle.

Figure 7.2 Sine = opposite/hypotenuse

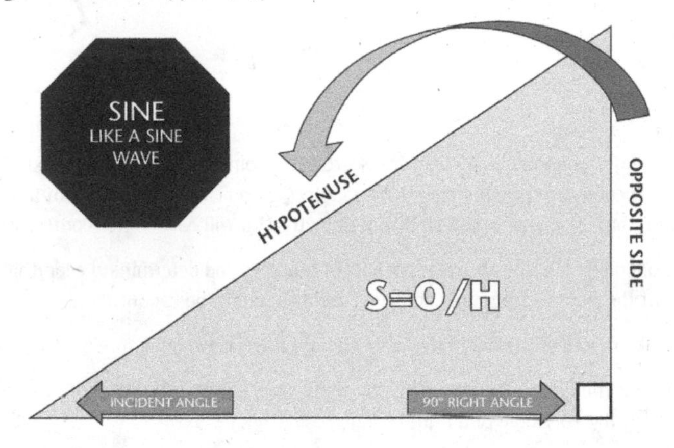

Figure 7.3 Cosine = adjacent/hypotenuse

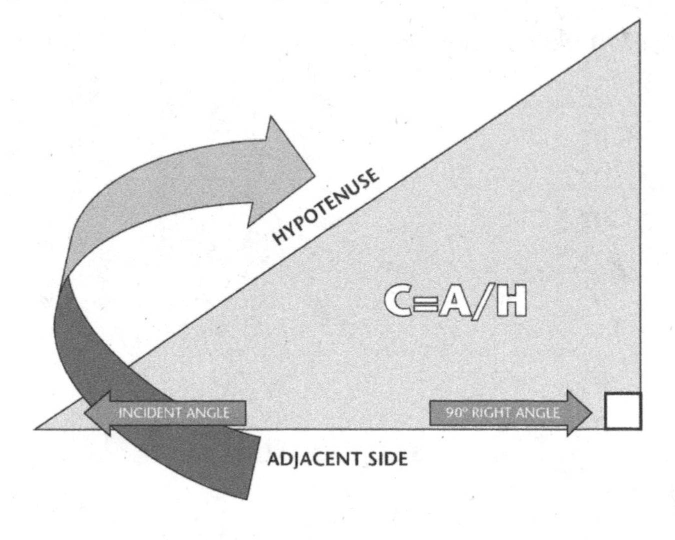

Figure 7.4 Tangent = opposite/adjacent

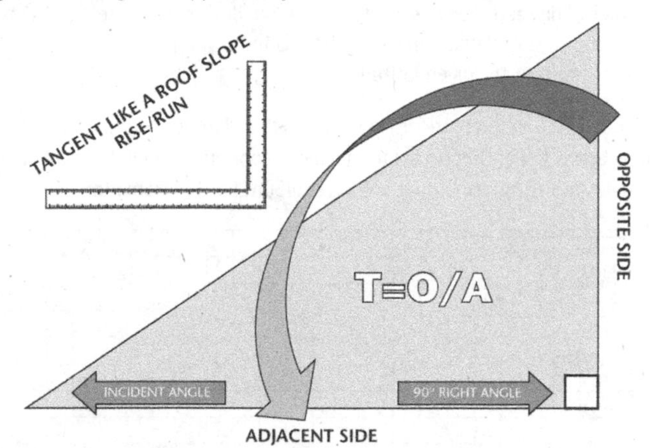

Common ways of remembering SOH CAH TOA are:

> Some Of Her
> Children Are Having
> Trouble Over Algebra

or:

> Some Old Hippie
> Caught Another Hippie
> Tripping On Acid

The crazier the saying, the easier it is to remember.

Now we will put it to use. With solar inter-row shading, we have an exercise where we can do all three functions in one exercise.

DETERMINING INTER-ROW SHADING FOR A SOUTH-FACING ARRAY

We need to know a few things first:

1. Module width (assuming **landscape** mounting orientation): **40"** for this example.

2. Module tilt angle: we will use **30°** for this example.
3. Time of day and year we wish to have no shading: we will use 9 a.m.–3 p.m. winter solstice ("shade free from 9–3" is the saying).
4. Sun chart for the given latitude.

Figure 7.5 Sun path chart for Newark, NJ, 40.7° latitude Sun path chart created by University of Oregon Solar Radiation Monitoring Laboratory Online Sun Path Calculator. http://solardat.uoregon.edu/SunChartProgram.html

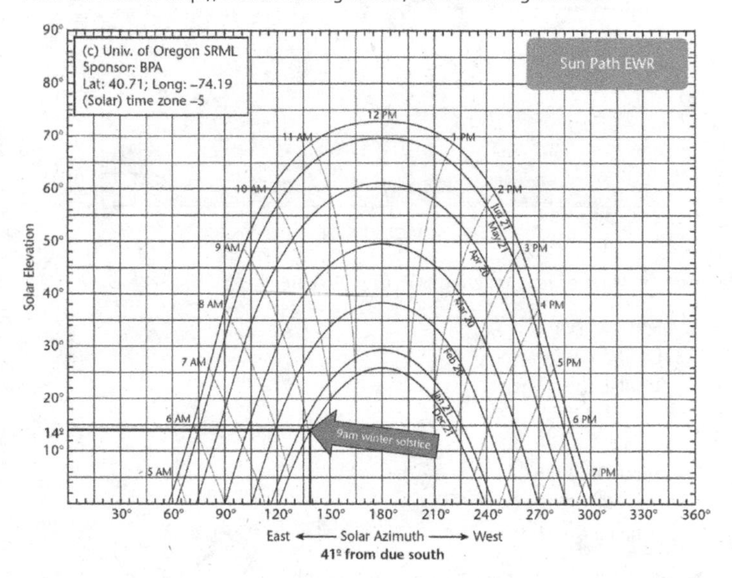

Look at the sun path chart and determine the solar elevation angle at 9 a.m. on the winter solstice, which is **14°**.

Determine the difference in azimuth from due south at 9 a.m. on the winter solstice, which is **41°**.

Now that we have all of the data, let's look at some right triangles.

In the first triangle, we have the **30° tilt angle** and the **hypotenuse** (which is the width of the module) and we need to find the height of the top of the array, which is the **opposite** side of the triangle from the angle.

Figure 7.6 PV module tilt and solar elevation triangles

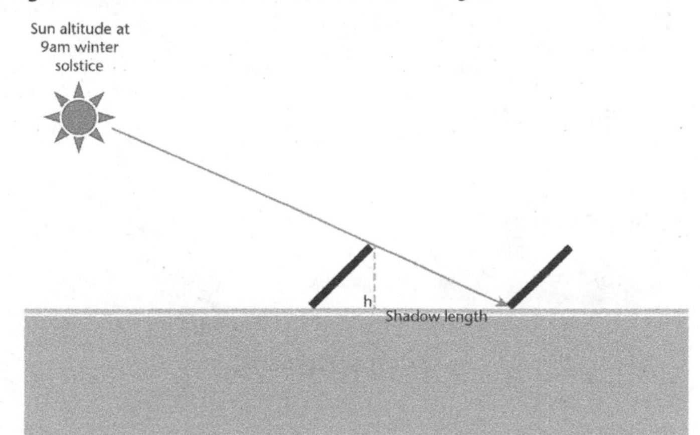

Figure 7.7 PV module tilt angle triangle

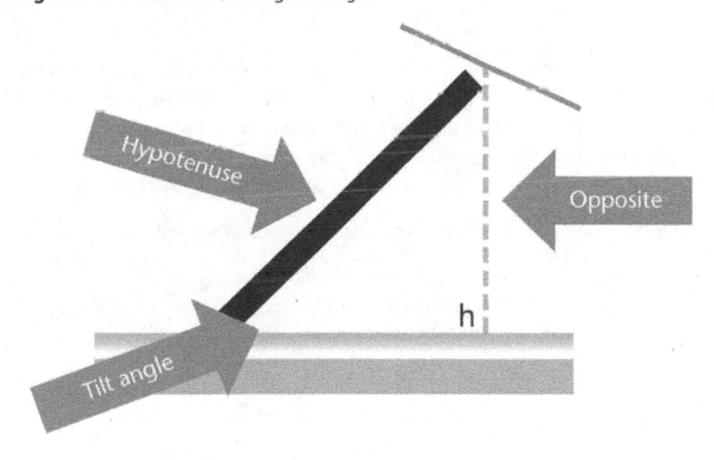

Out of SOH CAH TOA, we have O and H, so the function we will use is **SOH**, or:

Sine = opposite/hypotenuse
Opposite = sine × hypotenuse

Opposite = height = sine 30 × 40"
Height = 0.5 × 40" = 20"

Therefore the height is 20"

Figure 7.8 Solar elevation angle triangle

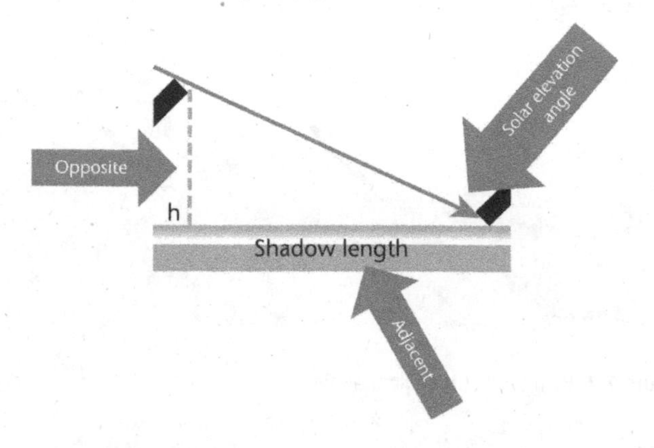

In the second right triangle, we have the **14° solar elevation angle** and the **opposite** side of the triangle, which is the height. We need to determine the length of the shadow at 9 a.m., which is the **adjacent** side of the triangle. If we did the calculation for 3 p.m., it would be the same, since the sun path is symmetrical. Both 9 a.m. and 3 p.m. are three hours from solar noon.

Out of SOH CAH TOA, we have O and A, so the function we will use is **TOA**, or:

Tangent = opposite/adjacent
Adjacent × tangent = opposite
Adjacent = opposite/tangent

Adjacent = shadow length = 20"/tan 14°
Shadow length = 20"/0.25 = 80"

Therefore the shadow length at 9 a.m. is 80"

The next triangle we will look for is from above as we see that the sunbeams are coming from the southeast direction at 9 a.m. in the morning, although we are lining up our array to face south.

Figure 7.9 Aerial view of azimuth correction angle from 9 a.m. to solar noon

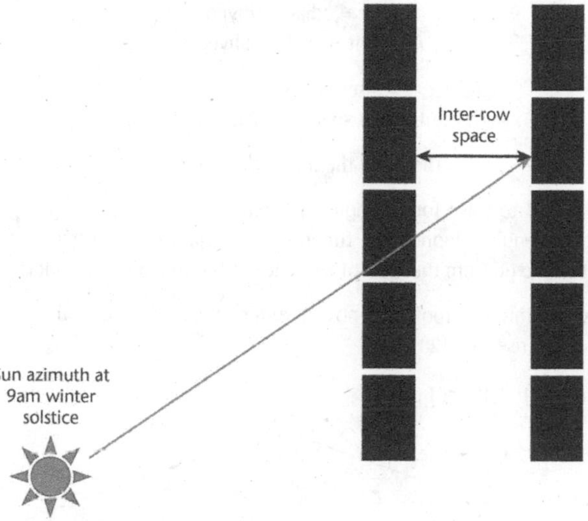

Figure 7.10 Azimuth correction angle triangle

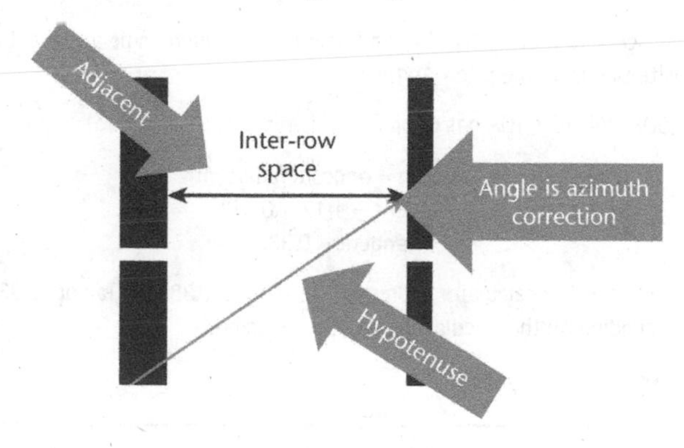

In the third right triangle, we have the **41° azimuth angle** correction and the **hypotenuse**, which is the shadow length at 9 a.m. We need to find the inter-row space, which is the missing **adjacent** side of the triangle.

Out of SOH CAH TOA, we have A and H, so the function we will use is **CAH**, or:

Cosine = adjacent/hypotenuse
Adjacent = cosine × hypotenuse

Adjacent = inter-row space = cosine 41 × 80"
Inter-row space = 0.755 × 80" = 60"

Therefore the inter-row space is 60"!

If you need to solve for an angle, rather than a side of a triangle, instead of using the regular trigonometry functions, you will use the inverse trigonometry functions to go from the ratio of the sides of the triangle to an angle.

We will do this for a roof slope now, assuming the roof slope is what roofers call 4:12 or a 4 rise to a 12 run.

Figure 7.11 4:12 roof pitch

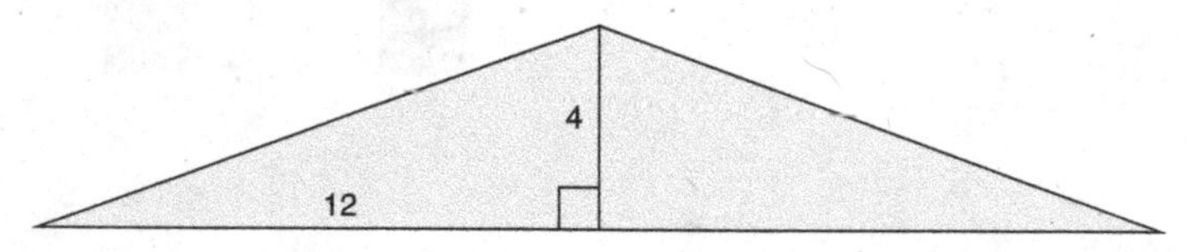

We can see here that the 12 is the **adjacent** to the pitch angle and the 4 is the **opposite** side from the **pitch angle**.

From SOH CAH TOA, TOA has opposite and adjacent:

Tangent = opposite/adjacent
Tangent = 4/12 = 0.333
Tangent = 0.333

What we do on the calculator is press 0.333 Tan^{-1} or 0.333 inv Tan or 0.333 shift Tan, depending on the calculator that you are using.

The answer is an 18° angle.

> At one of my first solar jobs, I left trigonometry all over my desk. Others were intimidated and I had extra job security. Trigonometry has a lot of bang for its buck when it comes to demonstrating intelligence.

Practice exam 1

70 questions, four hours

This practice exams in this book are by far the most valuable part of this book for those of you wanting to be NABCEP Certified. Once you get the basics down, the most efficient way to study for an exam is by taking practice exams and going over the answers in detail in the following chapter. Much of the basic concepts that we use are first introduced in the *Solar Photovoltaic Basics* book. Here we are putting those basic concepts in an advanced, realistic context. This book contains two full-length 70-question practice exams and a bonus exam, followed by detailed answers and explanations.

Set your timer for 4 hours. Sharpen your pencil. Use your NEC Codebook.

Ready, set, go!

1. What is the longest distance that electrical metallic tubing can go between supports?

- **a.** 10 feet
- **b.** 8 feet
- **c.** 4 feet
- **d.** 20 feet

2. How many stranded 10 AWG USE-2/RHW-2 conductors will fit in 2" EMT?

- **a.** 18
- **b.** 30
- **c.** 8
- **d.** 4

3. On a 4-wire, delta-connected system where the midpoint of one phase winding is grounded, what color should the ungrounded conductor with the highest voltage to ground be? (This is also known as the high leg.)

a. Orange
b. Red
c. Blue
d. Black

4. If you have 15 250W microinverters on a cable at 240V and a frequency of 60 Hertz, then what would be the minimum copper wire size for the 90C rated cable in a location with a high design temperature of 28C?

a. 8 AWG
b. 14 AWG
c. 10 AWG
d. 12 AWG

5. When installing PV source circuits in a rigid PVC conduit outside a building what should be done regarding an equipment grounding conductor?

a. Equipment grounding conductor is not required.
b. Equipment grounding conductor should be run inside the rigid PVC.
c. Bare copper equipment grounding conductor should be run on the outside of the rigid PVC.
d. Insulated equipment grounding conductor should be run on the outside of the rigid PVC.

6. What is the required depth and width of working space in front of a 8" wide 120/240Vac disconnect?

a. 3 feet 6", 8" width
b. 3 feet depth, 30" width
c. 30", 30" width
d. 3 feet depth, 3 feet width

7. An older inverter breaks and you replace it with a newer inverter. What must you check before replacing the inverter?

a. Make sure that the inverter MPP voltage window matches the voltage of the PV source circuit.
b. Make sure that you do not replace a grounded inverter with an ungrounded inverter.
c. Make sure that you do not replace an ungrounded inverter with a grounded inverter.

   d. Check to see if the new inverter does not have dc arc-fault protection if you are replacing an inverter without dc arc-fault protection.

8. An arc-flash hazard warning should be placed on equipment that may require servicing
   a. Within 7" of equipment
   b. While energized
   c. Below dc disconnect label
   d. By qualified persons

9. You are using typical 60 cell framed polycrystalline modules and notice at the worksite that the instructions that came with the modules in the pallet only have instructions for installing with mounting holes and you have a design for typical top clamps that are UL 2703 listed. What should you do?
   a. Install modules with top clamps only after getting written permission from top clamp manufacturer.
   b. Install modules with both top clamps and mounting holes.
   c. Install with top clamps that are UL 2703 listed on site.
   d. Only install modules according to module instructions.

10. Which of the following is an example of a dead load?
   a. HVAC
   b. Snow
   c. Wind
   d. Solar installer

11. The ASHRAE is important in the solar industry because, among other things
   a. ASHRAE has important data for firefighters in determining how to put out fires on buildings with solar and energy storage installations.
   b. ASHRAE has insolation data, which determines viability for solar installations.
   c. Local building codes adopt ASHRAE data for determining wind speed.
   d. NREL uses ASHRAE data for calculating kWh/kWp/yr.

12. Which is the most important for the AHJ (building department) when deciding to approve a permit?

a. 200kW PV system fire pathways
b. Interconnection agreements
c. Utility transformer size
d. Energy forecast

13. In an east-facing roof area with an installable area of 20 feet 6" x 11 feet 9" (20 feet 6" being the north to south dimension), how many 39" x 66" modules fit portrait and landscape? Assume 1" spaces between modules.

a. 12 portrait, 9 landscape
b. 12 portrait, 12 landscape
c. 9 portrait, 9 landscape
d. 15 portrait, 12 landscape

14. A tree is directly south of your solar array and is 20 feet tall. You are at 38° latitude and there is a flat rooftop that is 25 feet high. The closest part of the tree to the solar array is 30 feet away. If the tree will grow 2 feet per year, how long until the tree is shading the array?

a. 4.5 years
b. 16.3 years
c. 10.5 years
d. 8.15 years

15. In a seaside community, what is the best way of the following to attach a PV system to a composition asphalt shingle roof?

a. Brass screws into sheathing
b. Galvanized steel hangar bolts with a socket wrench
c. Zinc-plated bolts with an adjustable setting impact wrench
d. Stainless steel lag bolts with a torque wrench

16. If you are going to bury dc-to-dc converter source circuits under a parking lot, how deep must you bury the circuits that are in rigid metal conduit?

a. 36"
b. 24"
c. 12"
d. 18"

17. When you increase the size of a conductor cross-sectional area by 33% for voltage drop, how much must you increase the size of the corresponding equipment grounding conductor?

a. 16.5%
b. 50%
c. 33%
d. 0%

18. You are using an inverter that allows for energy storage backup and the backup circuit on a house is 120V. In taking the 120V backup circuit and putting it on a panelboard that is rated for 120/240V, you put the two busbars (L1 and L2) together to make the panelboard ready for the 120V circuit. You then run the circuits that you are going to back up to the new panelboard for the backed up loads. What should you do with the multi-wire branch circuits that you are going to back up?

a. Pull multiwire branch circuits through the entrance to the panelboard carefully.
b. Upsize the multiwire branch circuits neutral wire.
c. Remove the multiwire branch circuits.
d. Put up a sign indication that there are 120V multiwire branch circuits.

19. Your PV system will initially have a 83% derating when converting from kWh/square meter per day to kWh/kWp/day and on top of that, since you are tilted at latitude plus 15°, there is an additional 10% loss. How many kW of PV is required to offset 70% of your use if you use 22,000 kWh per year in a location with a daily average insolation of 5.3kWh/square meter/day?

a. 10.7kW
b. 5.5kW
c. 22.9kW
d. 3.1kW

20. Which of the following is NOT true about equalizing batteries?

a. Equalizing is used to reduce stratification in many flooded lead-acid batteries.

b. Equalizing is used to reduce stratification in sealed flooded lead-acid batteries.
c. Equalizing is done by increasing voltage.
d. Equalizing is done to prevent lead sulfate crystals from building up on the battery plates.

21. In which of the following cases is PV array dc ground fault protection not required?

a. A single module using a microinverter on a house
b. In areas without lightening if there is an additional auxiliary grounding electrode
c. On one and two family dwellings
d. Two strings on a solidly grounded ground mounted array used to pump water

22. Which of the following should be part of a grounding electrode system if it is present at the building?

a. Metal fences and fence posts
b. Metal water pipes
c. Metal flagpole
d. Metal gas pipe

23. Of the following, which is the smallest size conduit that can be used for 22 10 AWG USE-2/RHW-2 PV source circuits inside an attic?

a. 1" EMT
b. 2" EMT
c. 2.5" EMT
d. 1.5" EMT

24. You are called to a jobsite to look at an inverter and you measure the voltage from positive to ground and it is 0V and voltage from negative to ground is 408V. Which of the following could be correct?

a. It is a solidly grounded inverter with a loose connection from positive to negative.
b. It is a non-isolated inverter working properly.
c. It could be a positive grounded system or have a positive ground fault.
d. There is a positive to negative short circuit.

25. You have inverter detected ground faults that show up more on hot days. What could be a possible explanation?

- **a.** Pinched wire between module frame and rack
- **b.** Inverter output circuit wire insulation flaw
- **c.** Bad MC4 connector
- **d.** Blown string fuse

26. If you detect an intermittent ground fault, what would be the best course of action?

- **a.** Insolation testing
- **b.** Replace cable ties
- **c.** Insulation testing
- **d.** Upsize equipment grounding conductor

27. During an O&M call, you find a string that has slightly higher current than the other two strings on the inverter. The inverter has three strings going to a single MPP. You then turn the inverter off and the string that had higher current appears to still have higher current and the other two strings shut off. You are also looking at the monitoring and notice that about 3 weeks earlier the inverter started making about 35% less energy that would be expected. What is your diagnosis of this situation and what might you expect to find?

- **a.** There are likely two ground faults on the strings without the higher currents.
- **b.** There must be a ground fault on the string with the higher current.
- **c.** The string with the higher current has a positive to negative fault and I would expect to find a blown fuse in a combiner.
- **d.** There may be a bad connector or a connector that is not making contact, leaving one string not producing power.

28. What is the purpose of 690.12: rapid shutdown?

- **a.** Preventing shock
- **b.** Firefighter safety
- **c.** Employee safety
- **d.** Preventing fires

29. You are changing an old fuse grounded inverter with a new non-isolated inverter. The old inverter has USE-2 wire on the modules and a single fuse on the positive in the combiner. What must you do?

- **a.** You cannot install the new inverter if there is USE-2 wire on the modules.
- **b.** You must install fuses on both positive and negative polarities.
- **c.** Replace the inverter, keep the USE-2 wire, and make sure the disconnect opens both positive and negative.
- **d.** Replace the inverter with a fuse grounded inverter, which is easy to get.

30. For a free fall of how many feet is fall protection required?

- **a.** 10
- **b.** 6
- **c.** 12
- **d.** 6.5

31. According to OSHA, heavy equipment operators should

- **a.** Always wear a personal fall arrest system
- **b.** Be trained and certified by their employer
- **c.** Be trained and certified by OSHA
- **d.** Have a heavy equipment license

32. The arc-fault detection light is blinking on an inverter. You find that the connectors appear melted between two modules. What is the most likely problem?

- **a.** Squirrel living under array
- **b.** Improperly installed connectors
- **c.** Cold temperatures below -40C
- **d.** Snow and rain

33. Bad insulation around a wire with cracks would likely lead to?

- **a.** Voltage drop
- **b.** Ground fault
- **c.** Decrease in ampacity
- **d.** Increased insulation resistance

34. Battery storage systems shall have, according to the NEC

- **a.** Label with weight of the batteries
- **b.** Illumination not controlled by automatic means only
- **c.** Goggles on site at all times
- **d.** An energy storage maintenance logbook on site at all times out together by a qualified person

35. Which is 3-phase 240 delta with a high leg wire colors?

- **a.** Black, Red, Blue, White, Green
- **b.** Black, Red, White, Green
- **c.** Black, Orange, Blue, White, Green
- **d.** Brown, Orange, Yellow, Gray

36. What are the usual wire colors for 277/480 3-phase power?

- **a.** Black, Red, Blue, White
- **b.** Black, Orange, Yellow, Gray
- **c.** Brown, Orange, Yellow, Gray
- **d.** Black, Orange, Yellow, Black

37. At what point would rigid PVC require expansion joints if expansion is over?

- **a.** 1/4" or greater
- **b.** 1/2" or greater
- **c.** 3/4" or greater
- **d.** 1" or greater

38. Which type of electrode is unacceptable?

- **a.** Concrete encased electrode
- **b.** Steel frame of a building
- **c.** Copper wire buried around house
- **d.** Aluminum ground rod

39. What is the required width of working space in front of a PV system ac disconnecting means?

- **a.** The width of the ac disconnecting means or 30", whichever is less
- **b.** 30"

c. The width of the ac disconnect or 30", whichever is greater
d. The width of the equipment

40. Which of the following is typically the best solution for making wires inaccessible on a ground mount?

a. Security guards
b. The neighbor's electric fence
c. Covering over the backs of modules
d. Alarm system

41. What is the smallest allowable equipment grounding conductor going in conduit with 17 PV source circuits each with a short circuit current of 8A going in a single conduit to a combiner?

a. 14 AWG copper
b. 12 AWG copper
c. 6 AWG copper
d. 10 AWG copper

42. What is the UL listing for inverters?

a. 1703
b. 1741
c. 2703
d. 2741

43. In which situation is a bonding jumper not required when connecting EMT to a junction box in a system with over 250V?

a. When the EMT is aluminum
b. When using a WEEB
c. When there are ringed knockouts
d. When there are no concentric knockouts

44. What is NOT an example of when an interactive inverter would be operating off the MPPT point of the IV curve?

a. Clipping
b. Overheating
c. STC operation
d. Reducing power output for a non-exporting system

45. Which is the best tool for checking for intermittent ground faults?

- **a.** Digital multimeter
- **b.** Clamp on ammeter
- **c.** Megohmmeter
- **d.** Two alligator clips and a jumper

46. PV system disconnects in different locations than the main service disconnect will need what?

- **a.** OSHA Certification
- **b.** A plaque or directory
- **c.** UL 1703 listing
- **d.** IEC label

47. At what point must conductors have the ability to be rapidly shut down to less than 30V in less than 30 seconds outside of a building?

- **a.** At the module level
- **b.** 10 feet from the array
- **c.** 3 feet from the array
- **d.** 1 foot from the array

48. At 33° latitude with a latitude −15° tilt and peak sun hours at that tilt being 4.5, what would be the minimum kW system size needed to produce 75% of the energy if you considered a derating factor of 77%? Assume that the required amount of energy to be used is 9000kWh per year.

- **a.** 5.3kW
- **b.** 7.1kW
- **c.** 5.6kW
- **d.** 11.2kW

49. Battery room doors must?

- **a.** Open in the direction of egress
- **b.** Never be locked
- **c.** Always be locked
- **d.** Have ventilation built into the door

50. Which is NOT required by the NEC for a PV array installation?

- **a.** Grounding of 2703 listed rail systems

- **b.** Lightning protection
- **c.** Torque wrench as required by manufacturers' instructions
- **d.** Module frames that are at ground voltage

51. Given a 225A busbar and a 200A main breaker, what is the most current an inverter can have assuming a load-side, center-fed busbar?

- **a.** 32A
- **b.** 56A
- **c.** 70A
- **d.** 25A

52. Given the following voltage measurements of a 3-phase 4 wire system, which is the high leg?

- **a.** Phase A to neutral 120V
- **b.** Phase A to phase C 240V
- **c.** Phase C to neutral 120V
- **d.** Phase B to neutral 208V

53. How do you achieve desired torque settings for connectors?

- **a.** Using an impact driver
- **b.** Using a calibrated torque tool
- **c.** Using a differential metered torque sensor
- **d.** Using a trained wrist

54. You would like to produce 80% of the energy at a house with an annual consumption of 7500kWh per year. The slope of the roof is 5:12 and is facing 180 azimuth at latitude 38. With 82% system efficiency, what would be the required amount of PV? (Use the following information.)

Solar insolation and tilt for location at latitude 38
Tilt 0 = 4.9 PSH
Latitude - 15 = 5.2 PSH
Latitude = 5.4 PSH
Latitude + 15 = 5.1 PSH

- **a.** 3.9kW
- **b.** 4.1kW
- **c.** 3.1kW
- **d.** 5.0kW

55. What type of inverter/array grounding system includes fuse grounded inverters and most inverters without transformers?

- **a.** Functional grounded system
- **b.** Ungrounded system
- **c.** Isolated grounding system
- **d.** Non-isolated grounding system

56. When troubleshooting four strings, you notice that three strings have slightly less current than the fourth string, but that when you measure voltage, the fourth string has 0V and the other three strings have normally expected voltage. What is the likely problem?

- **a.** Ground fault on the fourth string
- **b.** Short circuit on the fourth string
- **c.** Fourth string disconnected
- **d.** Forth string open circuited

57. Using standard crystalline modules with 6" solar cells (156mm), what would be the most number of modules that you could parallel to a charge controller without PV source circuit fusing?

- **a.** 0
- **b.** 3
- **c.** 1
- **d.** 2

58. If a thin film PV module has a short circuit current of 1.8A and a maximum series fuse rating of 15A, what would be the maximum number of PV source circuits combined at a dc combiner without needing fusing?

- **a.** 2
- **b.** 8
- **c.** 1
- **d.** 3

59. Array operating current is 80% of expected on a PV output circuit. Using a clamp ammeter, all of the PV source circuits measure the same current except for one string which measures0A. The installer then pulls all of the fuses and does continuity tests on all of the fuses and they all check out good. What is the most likely scenario?

a. Shorted bypass diode
b. Module wired with reverse polarity
c. Disconnected wire
d. Intermittently blown fuse

60. Two PV source circuits of 10 220W 60 cell PV modules in series are connected to a single MPP inverter that was installed in 2008. You were asked to check on production and noticed that the array was shaded 1 foot from the edge on the short edge of most of one of the strings at noon in the winter. The modules were mounted in portrait. If you measured 1kW of production in this scenario, about what would be the production without the shaded strip running down the short edge?

a. 1kW
b. 1.5kW
c. 2kW
d. Greater than 2.5kW

61. A rooftop PV array is to be installed in a track home, where the HOA requires that no conduit be visible from outside of the building. Which of the following wiring methods will be permitted to run conduit in attic?

a. USE-2/RHW-2 in free air to junction box and then USE-2/RHW-2 in EMT
b. USE-2 in free air to junction box under modules and then USE-2 in EMT
c. USE-2 in free air to junction box under modules then THWN-2 in rigid NMC
d. USE-2/RHW-2 in free air to junction box under modules then THWN-2 in rigid NMC

62. What is the main reason for the inverter bypass switch in a utility-interactive, battery-backup system?

a. So that backed-up loads can operate during power outage
b. To charge batteries with the utility
c. To allow backed-up loads to be used during inverter failure or maintenance
d. So that anti-islanding can be controlled

63. 12 AWG USE-2/RHW-2 is lying on a tile roof and going 20 feet across the roof. What can be done to make this comply with the NEC?

- **a.** Run the cable in EMT.
- **b.** Install in Liquidtight Flexible NMC and secure every 6 feet.
- **c.** Support the cable on purlins.
- **d.** Put the cable under the tiles but above the roofing felt.

64. On a partly cloudy day an inverter intermittently shuts off for periods of 5 or 10 minutes at a time? What is the most likely cause of this problem?

- **a.** PV source circuit too far from inverter
- **b.** Undersized dc conductors
- **c.** Undersized ac conductors
- **d.** Intermittent ground fault

65. What would be the largest single phase inverter that you could put on a breaker in a 208V 125A subpanel with a 100A feeder breaker?

- **a.** 12kW
- **b.** 10.4kW
- **c.** 9600W
- **d.** 8320W

66. When connecting flexible, fine-stranded copper wire to an inverter, combiner box, or battery terminal lug, the installer must ensure that the terminal lug is rated for?

- **a.** Dc use only
- **b.** 75C
- **c.** Fine-stranded wire
- **d.** 90C

67. An interactive residential PV system is installed in Alaska on a single family home. The system is commissioned in July and the open circuit voltage measures 580V. If there is a utility failure in the winter, what could happen?

- **a.** The system would deliver more than normal power.
- **b.** The inverter warranty would likely be voided.
- **c.** The inverter would be likely to indicate a ground fault.
- **d.** The inverter would limit current.

68. If you install an 8 AWG grounding electrode conductor, which of the following is the best method?

a. In free air when not exposed to physical damage
b. In RMC, IMC, PVC, RTRC-XW, EMT, or cable armor
c. In USE-2 cable
d. Securely fastened to the surface of the building

69. What is the primary purpose of rapid shutdown for PV systems?

a. Preventing ground faults
b. Preventing fires
c. First responder safety
d. Preventing shocks for solar installers, electricians, roofers, and other construction workers

70. What is the maximum circuit current for a dc-to-dc converter source circuit?

a. Manufacturers' continuous output current rating
b. Isc
c. 125% of Isc
d. 156% of Isc

Practice exam 1

With answers and explanations

Chapter 9

This chapter is by far the most valuable part of this book. Once you get the basics down, the most efficient way to study for an exam is by taking practice exams and going over the answers in detail. Much of the basic concepts that we use are first introduced in the *Solar Photovoltaic Basics* book. Here we are putting those basic concepts in an advanced, realistic context. (The correct answer is emboldened.)

1. What is the longest distance that electrical metallic tubing can go between supports?

- **a. 10 feet**
- **b.** 8 feet
- **c.** 4 feet
- **d.** 20 feet

Look up EMT in the NEC in Article 358. 358.30(A) indicates that we cannot go further than 10 feet between supports.

Remember that all of the wiring methods, such as EMT are in Chapter 3: wiring methods and it should not be too hard to find. You can also look it up in the table of contents or index. Better yet, memorize 358 for the most popular PV wiring method in the US.

2. How many stranded 10 AWG USE-2/RHW-2 conductors will fit in 2" EMT?

- **a.** 18
- **b. 30**
- **c.** 8
- **d.** 4

There are different ways of determining the correct answer. There are Tables 1, 4, and 5 in Chapter 9 and then there is the shortcut in Informative Annex C.

Let's start with the easy method first and then do it the more difficult way.

First, we will note that the wire in question is dual listed, which means it can fulfill the requirements of both USE-2 and RHW-2. We can see that USE-2 is not in the tables for determining how many conductors fit in conduit and that is because USE-2 on its own cannot go in conduit, but when it is dual listed, it can and we will use RHW-2 for the rest of our problem solving here.

First way to solve:

In Annex C, Table C1, you can look at the top of the table and cross reference 10 AWG RHW-2 and 2" EMT and you will see that you can have 30 conductors in conduit maximum here. I imagine that would be a tight fit!

Annex C only works if all of the conductors are of the same size, but sometimes with a ground wire of a different size, the right answer will be obvious with Annex C.

Second way to solve:

With the Chapter 9 tables, let us go to Table 4: dimensions and percent area of conduit and tubing to see what we are dealing with and we will start with the "over 2 wires" column most of the time, since we are usually running over two wires in the conduit. (Chapter 9, Table 1 shows us that we are only allowed to physically fill up 40% of the space, which leaves 60% of the space for air and room for the wires to move in.) Here we can see that trade size 2 (2" EMT) has 40% of its area that we can use for wires as 1.342 in^2.

Next, we will go to Chapter 9, Table 5: dimensions of insulated conductors and fixture wires to figure out the cross sectional area of 10 AWG RHW-2. We can see that 10 AWG RHW-2 has a cross-sectional area of 0.0437 in^2.

Now we just do simple division to see how many fit:

$$1.342\ in^2 / 0.0437\ in^2 = 30.7$$

So the answer once again is we can fit 30 at most. Electricians never want to put the maximum, since it can be very difficult to work with. If you are a designer, be kind to the electrician, and give them some extra room in that conduit!

3. On a 4-wire, delta-connected system where the midpoint of one phase winding is grounded, what color should the ungrounded conductor with the highest voltage to ground be? (This is also known as the high leg.)

a. Orange
b. Red
c. Blue
d. Black

A high leg is common on 240/120V 3-phase systems and sometimes referred to as a stinger. At times electricians have measured a hot wire to ground and found 120V and thought they were dealing with 120V for all of the ungrounded conductors to ground, however, the high leg will be 208V and will often damage 120V loads that are connected to it. If you see orange, think not 120V.

4. If you have 15 250W microinverters on a cable at 240V and a frequency of 60 Hertz, then what would be the minimum copper wire size for the 90C rated cable in a location with a high design temperature of 28C?

a. 8 AWG
b. 14 AWG
c. 10 AWG
d. 12 AWG

First, calculate the current of each microinverter, which is:

250W / 240V = 1.04A

Then, multiply by number of microinverters on the cable:

1.04A x 15 inverters = 15.6A

Thoughts:

The temperature is under 30C, so no need to derate for ambient temperature here. I would also assume that, in this situation, the microinverter cable is going to a circuit breaker. Additionally, we will assume that this microinverter cable, like most is not in conduit. We will use Table 310.15(B)(16), since a microinverter cable will have multiple conductors in it and 310.15(B)(17) is for single insulated cable.

Multiply 15.6A x 1.25 get circuit breaker size = 19.5A.

19.5A rounds up to a 20A circuit breaker.

240.4(D) small conductor rule says that a 20A breaker should have at least a 12 AWG wire (assuming copper).

In Table 310.15(B)(16) we see that a 12 AWG 90C rated copper conductor can take 30A, so that is well over the requirement for this 19.5A cable.

5. When installing PV source circuits in a rigid PVC conduit outside a building what should be done regarding an equipment grounding conductor?

- **a.** Equipment grounding conductor is not required.
- **b.** **Equipment grounding conductor should be run inside the rigid PVC.**
- **c.** Bare copper equipment grounding conductor should be run on the outside of the rigid PVC.
- **d.** Insulated equipment grounding conductor should be run on the outside of the rigid PVC.

It is acceptable to use rigid PVC for PV source circuits on the outside of a building. We can find out about the wiring methods for rigid PVC in Chapter 3: wiring methods Article 352: rigid polyvinyl chloride conduit: type PVC.

It does not take too long to skim through Article 352 and we can see that 352.60: grounding tells us that the equipment grounding conductor should be run inside of the conduit. An equipment grounding conductor is required for PV systems on rooftops.

There are many other things we can easily look up in Article 352, such as coefficients of expansion, how far apart supports should be, and locations where we can install the conduit.

6. What is the required depth and width of working space in front of a 8" wide 120/240Vac disconnect?

- **a.** 3 feet 6" depth, 8" width
- **b.** **3 feet depth, 30" width**
- **c.** 30" depth, 30" width
- **d.** 3 feet depth, 3 feet width

110.26: spaces about electrical equipment is where to look and a good thing to memorize. For the depth of working space, we can look to Table 110.26(A)(1) and for a 120/240Vas system, we have 120V as voltage to ground, so under any condition, a required **depth of working space is 3 feet.**

110.26(A)(2): width of working space tells us that the width should be the **width of the equipment or 30", whatever is greater.**

While we are at it, the height of a working space around equipment should be at least 6½ feet or the height of the equipment, whichever is greater.

7. An older inverter breaks and you replace it with a newer inverter. What must you check before replacing the inverter?

- **a.** **Make sure that the inverter MPP voltage window matches the voltage of the PV source circuit.**
- **b.** Make sure that you do not replace a grounded inverter with an ungrounded inverter.
- **c.** Make sure that you do not replace an ungrounded inverter with a grounded inverter.
- **d.** Check to see if the new inverter does not have dc arc-fault protection if you are replacing an inverter without dc arc-fault protection.

The best answer here is to check that the inverter MPP voltage window matches the voltage of the PV source circuit. This can be done with string sizing, to make sure that the inverter will stay on during a hot day. Additionally, we should check for the maximum voltage to make sure that the inverter voltage does not go over the voltage limit of the inverter on a cold day.

For the wrong answers here, just think that the Code is probably not going to prevent our doing something safer, such as replacing a grounded inverter with an ungrounded inverter or installing an inverter with dc arc-fault protection.

8. An arc-flash hazard warning should be placed on equipment that may require servicing

- **a.** Within 7" of equipment
- **b.** **While energized**
- **c.** Below dc disconnect label
- **d.** By qualified persons

The best answer can be found in 110.16(A): arc-flash hazard warning. Arc flashes are more likely when working on energized equipment.

Arc-flash hazard warning labels should be put on equipment such as anything that requires:

- **Examination**
- **Adjustment**
- **Servicing**
- **Maintenance**

Arc flashes are most likely on equipment that is subject to higher available currents.

9. You are using typical 60 cell framed polycrystalline modules and notice at the worksite that the instructions that came with the modules in the pallet only have instructions for installing with mounting holes and you have a design for typical top clamps that are UL 2703 listed. What should you do?

a. Install modules with top clamps only after getting written permission from top clamp manufacturer.
b. Install modules with both top clamps and mounting holes.
c. Install with top clamps that are UL 2703 listed on site.
d. **Only install modules according to module instructions.**

You must always follow instructions. In the NEC, following instructions is covered in 110.3(B). Although the top clamp system may be better and more widely used, if the modules were not UL listed and tested with top clamps, then you cannot use them. If you installed with both top clamps and mounting holes, that would definitely not be how they were listed and tested.

10. Which of the following is an example of a dead load?

a. **HVAC**
b. Snow
c. Wind
d. Solar installer

A dead load is something that is permanent, such as HVAC, a solar array, or roofing materials. Temporary things, such as workers and snow are considered a live load. Wind is a wind load.

11. The ASHRAE is important in the solar industry because, among other things

- **a.** ASHRAE has important data for firefighters in determining how to put out fires on buildings with solar and energy storage installations.
- **b.** ASHRAE has insolation data, which determines viability for solar installations.
- **c. Local building codes adopt ASHRAE data for determining wind speed.**
- **d.** NREL uses ASHRAE data for calculating kWh/kWp/yr.

American Society of Heating Refrigeration Air-conditioning Engineers (ASHRAE) collects data about weather that we often use when designing solar systems. We can determine wind data, temperature data, including low and high temperatures. Most often installers are used to getting the ASHRAE design temperatures for calculating string sizing and determining ambient temperatures. Just think of information that would help someone designing heating, refrigeration, and air-conditioning systems and that information is put together by ASHRAE. It is interesting, that although ASHRAE is "American" the data are also compiled for other parts of the world.

12. Which is the most important for the AHJ (building department) when deciding to approve a permit?

- **a. 200kW PV system fire pathways**
- **b.** Interconnection agreements
- **c.** Utility transformer size
- **d.** Energy forecast

The AHJ (building department) should be mostly concerned about fires and safety, so fire pathways on a large system would be very important. Utility aspects of a system, such as interconnection agreements and utility transformers, are also important, but are more the domain of the utility. Production forecasting is important, too, but the AHJ usually does not fret about performance, since it is not their job.

13. In an east-facing roof area with an installable area of 20 feet 6" x 11 feet 9" (20 feet 6" being the north to south dimension), how many 39" x 66" modules fit portrait and landscape? Assume 1" spaces between modules.

- **a. 12 portrait, 9 landscape**
- **b.** 12 portrait, 12 landscape
- **c.** 9 portrait, 9 landscape
- **d.** 15 portrait, 12 landscape

Since this is an east-facing array, then the north to south dimension is the width of the array. Since it says an "installable area," we are going to assume that we do not need to account for fire setbacks.

With the 1" space between modules, we will have one less space than there are number of modules, because you do not need a space at the end. For example, if you had three modules, you would need only two spaces between modules.

This type of calculation is not difficult, but is time consuming when done within a timed test, and this can lead to mistakes.

Portrait calculations

11 feet 9" is the ridge to gutter dimension, which we need to convert to inches:

11 feet x 12"/ft = 132" = 11 feet
11 feet 9" = 132" + 9" = 141"

For the next part, I am going to make it simple and have an extra inch left over and then double check to make sure that that extra inch did not make a difference:

Length of the module = 66".

Space in between is 1" (there will be an extra inch left over here):

66" + 1" = 67"
141"/67" = 2.1 modules

Now to confirm two modules:

(66" x two modules) + (1" space) = 133"

So, we will only need 133" and could have used 141" portrait.

Next, we will see how many fit width wise (north to south) using the same method.

20 feet 6" is the width dimension of the roof, which we need to convert to inches:

20' x 12"/ft = 240" = 20 feet
20'6" = 240" + 6" = 246"

For the next part, I am going to make it simple and have an extra inch left over and then double check to make sure that that extra inch did not make a difference.

Width of the module = 39".

Space in between is 1" (there will be an extra inch left over here):

39" + 1" = 40"
246"/40" = 6.15 modules

Now to confirm six modules:

(39" x six modules) + (5" space) = 239"

So, we will only need 239" and could have used 246" portrait.

So for a portrait installation, we would be able to fit 2 x 6 = 12 modules.

Landscape calculations

Previously we already calculated the ridge to gutter dimension of 141".

For the next part, I am going to make it simple and have an extra inch left over and then double check to make sure that that extra inch did not make a difference.

Width of the module – 39".

Space in between is 1" (there will be an extra inch left over here):

39" + 1" = 40"
141"/40" = 3.525 modules

Now to confirm three modules:

(39" x three modules) + (2" space) = 119"

So, we will only need 119" and could have used 141" landscape.

Next we will see how many fit width wise (north to south) using the same method.

Previously we calculated the width of the roof is 246".

For the next part, I am going to make it simple and have an extra inch left over and then double check to make sure that that extra inch did not make a difference.

Length of the module = 66".

Space in between is 1" (there will be an extra inch left over here):

66" + 1" = 67"
246"/67" = 3.67 modules

Now to confirm three modules:

(66" x three modules) + (2" space) = 200"

So, we will only need 200" and could have used 246" portrait.

So for a portrait installation, we would be able to fit 3 x 3 = 9 modules.

For this type of long and easy calculation, you may wish to do it at the end of the exam, so you are not as pressed for time. You can see how you are progressing against the clock and do what feels right. Many people finish the exam early, so that would be a good reason to answer the question right away. On the other side of the coin, rushing through a question like this at the beginning of an exam may cause some people to panic and throw them off their game.

14. A tree is directly south of your solar array and is 20 feet tall. You are at 38° latitude and there is a flat rooftop that is 25 feet high. The closest part of the tree to the solar array is 30 feet away. If the tree will grow 2 feet per year, how long until the tree is shading the array?

- **a.** 4.5 years
- **b.** 16.3 years
- **c.** **10.5 years**
- **d.** 8.15 years

This is a trigonometry question. If you have no idea about trigonometry, you can still take a guess and still pass the exam. Many people are trigophobic.

The plan here is to figure out how far above current tree height the tree has to be to shade the PV and then to determine the years of growth at a rate of 2 feet per year we have to wait until the tree is a problem.

We need to figure out the angle of the sun above the horizon at winter solstice at solar noon due south. We are at 38° latitude for this question. At 38° latitude, we would need to have a tilt angle of 38° to face the sun at equinox, when the sun is over the equator. At winter solstice, we would have to tilt the module about 23.5° more to face the sun, since the tilt of the earth is about 23.5°:

38° degrees + 23.5° = 61.5° tilt

If we are tilted at 90° degrees, then we would be tilting at the horizon and we want to determine how far above the horizon the sun will be at winter solstice. We can subtract our tilt angle, which would be how far from overhead the sun would be from 90 to get the elevation angle of the sun. (The angle from overhead is called the zenith angle by star watchers):

90° - 61.5° = 28.5° elevation angle at winter solstice

We can see in the trigonometry chapter in this book, that if we know two things about a right triangle, we can then figure out the rest. One of the things we know here is the angle, which is 28.5° and another thing we know is a side of the triangle, which is the distance from the tree to the solar array and is 30 feet. Now what we need to solve for is the side of the triangle opposite of the 30 feet side, which is the height of the tree greater than the roof height.

From the trig chapter, if we know one side and an angle and need to know the opposite side, then we would use the tangent function (no hypotenuse needed). Since the sun is low and the angle is less than 45°, we can visualize that the 30 feet side of the triangle that we know is going to be greater than the opposite side that we are going to solve for. With a calculator find out what the tangent of our 28.5° angle is:

28.5 tangent = 0.543

0.543 is the ratio of the side we know to the side of the triangle we do not know, so multiply 0.543 by the side that we do know

0.543 x 30 feet = 16.3 feet above the height of the rooftop

Since the tree is currently 20 feet tall and the height of the roof is 25 feet tall, then the tree is 5 feet below the height of the roof, so we will add 5 feet to 16.3 feet to get:

16.3 feet + 5 feet = 21.3 feet of tree growth until shading

At a growth rate of 2 feet per year to grow 21.3 feet, we can divide 2 feet into 21.3 feet to get:

21.3 feet/2 feet per year = 10.65 years

So it would take about 10.5 years

15. In a seaside community, what is the best way of the following to attach a PV system to a composition asphalt shingle roof?

- **a.** Brass screws into sheathing
- **b.** Galvanized steel hangar bolts with a socket wrench
- **c.** Zinc-plated bolts with an adjustable setting impact wrench
- **d.** **Stainless steel lag bolts with a torque wrench**

Especially in a seaside location, which may be wet and where corrosion may speed up due to the moisture, it is recommended to use stainless steel hardware to fasten a solar system to a roof. Additionally, when installing the lag bolts and when installing many things, you are supposed to use a torque wrench. In practice, many people do not use a torque wrench when they should. You should even use a torque screw driver when you attach a wire to a circuit breaker.

In an overview of the wrong answers here, zinc plating is recommended for indoors, whereas hot-dip galvanizing is more appropriate for outdoors. Stainless steel is, however, superior to galvanized. You are likely to see some galvanized racks on utility scale PV projects, since it is cost saving and you are more likely to see galvanized farther away from the sea.

Brass is expensive and often used in marine applications. Attaching to sheathing is not as strong as attaching to rafters of a typical American house. That being said, there are racking systems that may be good that do attach to sheathing, but there are conflicting opinions on whether a sheathing attachment is sufficient. Sheathing is plywood or OSB (oriented strand board) and looks like particle board.

16. If you are going to bury dc-to-dc converter source circuits under a parking lot, how deep must you bury the circuits that are in rigid metal conduit?

- **a.** 36"
- **b.** **24"**
- **c.** 12"
- **d.** 18"

NEC Table 300.5: minimum cover requirements is where you will find the answer. Think of Chapter 3: wiring methods and materials and Article 300: general requirements for wiring methods and materials and you will have it covered. Without knowing to look at Table 300.5, you would just have to take a guess. You can also see here that every wiring method under a parking lot is 24".

17. When you increase the size of a conductor cross-sectional area by 33% for voltage drop, how much must you increase the size of the corresponding equipment grounding conductor?

- **a.** 16.5%
- **b.** 50%
- **c.** 33%
- **d.** **0%**

When upsizing a conductor for voltage drop, you do not need to upsize the equipment grounding conductor. This is covered in 690.45: size of equipment grounding conductors, which tells you to use Table 250.122, where you will size your equipment grounding conductor (EGC) based on the size of the overcurrent protection device.

On another note, when we are not required to have an overcurrent protection device, such as when we have one or two PV source circuits (strings) going to a single MPP, then 690.45 directs us to use 690.9(B), which is essentially telling us to use 125% of maximum circuit currents, which means Isc x 1.56 in place of the overcurrent protection device, which makes sense, since that is how we would determine an overcurrent protection device size.

18. You are using an inverter that allows for energy storage backup and the backup circuit on a house is 120V. In taking the 120V backup circuit and putting it on a panelboard that is rated for 120/240V, you put the two busbars (L1 and L2) together to make the panelboard ready for the 120V circuit. You then run the circuits that you are going to back up to the new panelboard for the backed up loads. What should you do with the multiwire branch circuits that you are going to back up?

- **a.** Pull multiwire branch circuits through the entrance to the panelboard carefully.
- **b.** Upsize the multiwire branch circuits neutral wire.
- **c.** **Remove the multiwire branch circuits.**
- **d.** Put up a sign indication that there are 120V multiwire branch circuits.

A multiwire branch circuit, in this case, is a conductor that has two different 120V circuits that share a neutral. When these circuits are on a 120/240V panelboard, the two different circuits will be out of phase with one another and the currents on the neutral will cancel one another out. If we change these circuits to being in phase with one another, then the neutral currents would add up rather than cancelling one another out. In the 2014 NEC, we would find the rule for this in Article 690, however, now it is in the new Article 710: stand-alone systems. Section 710.15(C): single 120-volt supply tells us that we can use 120/240V service equipment when there are no multiwire branch circuits. We would also need a sign saying: WARNING: SINGLE 120-VOLT SUPPLY. DO NOT CONNECT MULTIWIRE BRANCH CIRCUITS!

19. Your PV system will initially have a 83% derating when converting from kWh/square meter per day to kWh/kWp/day and on top of that, since you are tilted at latitude plus 15° degrees, there is an additional 10% loss. How many kW of PV is required to offset 70% of your use if you use 22,000 kWh per year in a location with a daily average insolation of 5.3kWh/square meter/day?

- **a.** **10.7kW**
- **b.** 5.5kW
- **c.** 22.9kW
- **d.** 3.1kW

If we want to offset 70% of 22,000kWh per year, we can first figure out what 70% of 22,000 is:

0.7 x 22,000kWh = 15,400kWh

Now we want to figure out how many kWh a single kW of PV will make in these conditions.

To figure out how many kWh a kW will make in a day, we can derate our insolation by 83%:

0.83 x 5.3kWh/square meter/day = 4.40kWh/kWp/day

Then we will have an additional 10% loss, which means we will keep 90% and our derating factor for a 10% loss will be 0.9:

4.4kWh/kWp/day x 0.9 = 3.96 kWh/kWp/day

We then need to figure out our annual kWh production for a single kW of PV installed, so:

3.96 kWh/kWp/day x 365 days per year = 1445 kWh per year

Since one kW of PV will make 1445 kWh per year and we need to make 15,400 kWh per year, we can just divide our kWh made by a kW into the number of kWs we need to make to get kWs required to install in order to make that much energy:

15,400 kWh per year/1445 kWh/kWp/yr = 10.7kW required

20. Which of the following is NOT true about equalizing batteries?

a. Equalizing is used to reduce stratification in many flooded lead-acid batteries.
b. **Equalizing is used to reduce stratification in sealed flooded lead acid batteries.**
c. Equalizing is done by increasing voltage.
d. Equalizing is done to prevent lead sulfate crystals from building up on the battery plates.

If there is a question that says which one is *not*, then that must mean that three out of four possible answers are true. Equalizing is done by increasing voltage and causing water molecules to be split into hydrogen and oxygen gas molecules in the form of bubbles. These bubbles stir up the battery acid, getting rid of stratification. Stratification is when the heavier molecules rest at the bottom of the battery cells. Additionally, the bubbles rising will scrape the lead sulfate crystals off of the lead plates. You typically only equalize batteries that you can add fluid to in order to make up for the loss in water molecules that are split into hydrogen and oxygen. We will add distilled water to the maintenance on these batteries to extend their life. A sealed battery is considered maintenance free and we are unable to do equalization on a sealed battery, since it would end up drying out prematurely.

On another note, having a charge setting too high will also dry out and wreck a lead-acid battery. In hot weather, the charge setting should be lower than when it is cold, which is why lead-acid batteries often have temperature sensors that communicate with charge controllers. As it is warmer, the chemical reactions speed up and batteries could dry up.

21. In which of the following cases is PV array dc ground fault protection not required?

- **a.** A single module using a microinverter on a house
- **b.** In areas without lightening if there is an additional auxiliary grounding electrode
- **c.** On one and two family dwellings
- **d. Two strings on a solidly grounded ground mounted array used to pump water**

According to 690.41(B): ground-fault protection exception:

> **PV arrays with not more than two PV source circuits and with all PV system dc circuits not on or in buildings shall be permitted without ground-fault protection where solidly grounded.**

Although there is this exception, it is not used by most installers, since solidly grounded PV systems are rare. One of the ways that a solidly grounded system *is* used would be a direct water pumping system. Most modern equipment, such as inverters, have ground fault protection.

22. Which of the following should be part of a grounding electrode system if it is present at the building?

- **a.** Metal fences and fence posts
- **b. Metal water pipes**
- **c.** Metal flagpole
- **d.** Metal gas pipe

Article 250: grounding and bonding. Part III: grounding electrode system and grounding electrode conductor is where you should start looking to learn about grounding electrodes. This Part starts with 250.50: grounding electrode system, where we are told that if a metal water pipe, metal in-ground support structure, concrete encased electrode, ground ring, other listed electrodes, and plate electrodes should all be connected as part of a grounding electrode system.

Gas pipes are not permitted to be part of the electrode system.

23. Of the following, which is the smallest size conduit that can be used for 22 10 AWG USE-2/RHW-2 PV source circuits inside an attic?

- **a.** 1" EMT
- **b.** 2" EMT

c. **2.5" EMT**
d. 1.5" EMT

Since all of the conductors are the same size, it would be easiest to use Informative Annex C, Table C.1, however, we could still have used Chapter 9, Tables 1, 4, and 5.

At the beginning of Table C.1 is RHW-2. USE-2 is not in the table. The wire is dual listed and we are using the RHW-2 for this exercise.

We look at 10AWG on the left side of the table and then go right until we find the right number of conductors. The trick here is that with 22 PV source circuits, we will have 44 conductors, a positive and a negative for each circuit. When we go right, we can see that 2" can hold 30 conductors and 2.5" can hold 53 conductors, so 2" is too small and 2.5" is the correct answer.

If there were an answer that was not a metal raceway, then that would not work, since PV source circuits in a building have to be in a metal raceway or MC cable.

24. You are called to a jobsite to look at an inverter and you measure the voltage from positive to ground and it is 0 volts and voltage from negative to ground is 408V. Which of the following could be correct?

a. It is a solidly grounded inverter with a loose connection from positive to negative.
b. It is a non-isolated inverter working properly.
c. **It could be a positive grounded system or have a positive ground fault.**
d. There is a positive to negative short circuit.

This could be a positively grounded system that is in perfect condition and it could also be a PV system that has a positive ground fault.

If there were a loose connection from positive to negative, it could be intermittent shorting and shorting itself would measure 0V from positive to negative.

A non-isolated inverter (formerly known as an ungrounded inverter in the 2014 NEC) would measure the same voltage to ground from negative or positive. Ground would be in the middle.

We are unlikely to see a new system that is positively grounded, however, there are many out there that we may be troubleshooting.

25. You have inverter detected ground faults that show up more on hot days. What could be a possible explanation?

- **a.** **Pinched wire between module frame and rack**
- **b.** Inverter output circuit wire insulation flaw
- **c.** Bad MC4 connector
- **d.** Blown string fuse

On a hot day, the insulation of a wire can be less likely to protect the wire and could lead to ground faults. It is common to have ground faults from pinched wires from sloppy installations with bad wire management. You could also get the same ground faults on wet days.

Inverters do not typically measure ground faults on inverter output circuits; they detect dc ground faults.

A bad MC4 connector is less likely to cause a ground fault than a pinched wire.

A blown string fuse would be a sign of a short circuit, not a ground fault.

26. If you detect an intermittent ground fault, what would be the best course of action?

- **a.** Insolation testing
- **b.** Replace cable ties
- **c.** **Insulation testing**
- **d.** Upsize equipment grounding conductor

On a modern inverter, the inverter will do a test that is very similar to an insulation test, which is like a quick pulse of voltage and then testing the resistance of the insulation by comparing the conductor to ground. If there is not enough resistance, the there is a problem with the insulation. A trade name that is often used for insulation testing is a Megger Test.

The insulation test would be the best course of action. Changing the inverter would not fix a ground fault. Upsizing the equipment grounding conductor would not fix a ground fault. Replacing cable ties would probably not help much either.

27. During an O&M call, you find a string that has slightly higher current than the other two strings on the inverter. The inverter has three strings going to a single MPP. You then turn the inverter off and the string that had higher current appears to still have higher current and the other two strings shut off. You are also looking at the monitoring and notice that about 3 weeks earlier the inverter started making about 35% less energy that would be expected. What is your diagnosis of this situation and what might you expect to find?

- **a.** There are likely two ground faults on the strings without the higher currents.
- **b.** There must be a ground fault on the string with the higher current.
- **c.** **The string with the higher current has a positive to negative fault and I would expect to find a blown fuse in a combiner.**
- **d.** There may be a bad connector or a connector that is not making contact, leaving one string not to produce power.

After a short circuit (fault) positive to negative of a string, it is likely that there will be a blown fuse and the shorted string would keep on shorting. It is also a situation where the inverter would keep on producing power with the strings that are not shorted out. In this situation, the shorted string would have slightly higher current, since it would be at Isc, rather than Imp.

Ground faults would make the inverter shut down and there would be no power expected. A bad connector would have no current, rather than slightly higher current.

28. What is the purpose of 690.12: rapid shutdown?

- **a.** Preventing shock
- **b.** **Firefighter safety**
- **c.** Employee safety
- **d.** Preventing fires

690.12: rapid shutdown of PV systems on buildings is for firefighter safety. Employees would have time to properly shut down a system to work on it and employees should already be educated on PV systems. It has been said that firefighters will avoid putting out fires on houses with PV, since they do not understand how they work and have heard that they can still have dangerous voltage after the power to the building has been cut off. With rapid shutdown, the firefighters know that they can read a sign, hit a switch, and see that the array is safe.

Another answer here is shock and it could be a good answer if firefighter safety were not on the list, since we do not want firefighters to be shocked, however, firefighter safety is the best answer.

Rapid shutdown was not intended to prevent fires; it was intended for firefighter safety.

29. You are changing an old fuse grounded inverter with a new non-isolated inverter. The old inverter has USE-2 wire on the modules and a single fuse on the positive in the combiner. What must you do?

- **a.** You cannot install the new inverter if there is USE-2 wire on the modules.
- **b.** You must install fuses on both positive and negative polarities.
- **c.** **Replace the inverter, keep the USE-2 wire, and make sure the disconnect opens both positive and negative.**
- **d.** Replace the inverter with a fuse grounded inverter, which is easy to get.

In the 2014 NEC days, the correct answer would have been a, however according to the 2017 NEC, the right answer is c. The 2014 NEC made it more difficult to replace the inverter with the safer inverter than with the less safe fuse grounded inverter. This is why we are now allowed to use USE-2 wire on "formerly known as ungrounded" non-isolated inverters. This also goes for the single polarity fuse. We do not want people to go looking for fuse grounded inverters or having to replace an entire solar array, because the older, less safe inverter broke.

30. For a free fall of how many feet is fall protection required?

- **a.** 10
- **b.** **6**
- **c.** 12
- **d.** 6.5

You should always know that fall protection is required at 6 feet!

31. According to OSHA, heavy equipment operators should

- **a.** Always wear a personal fall arrest system
- **b.** **Be trained and certified by their employer**
- **c.** Be trained and certified by OSHA
- **d.** Have a heavy equipment license

OSHA says that a heavy equipment operator should be trained and certified by their employer.

Fall protection is not required by most jobs. Fall protection is required where someone may fall 6 feet.

32. The arc-fault detection light is blinking on an inverter. You find that the connectors appear melted between two modules. What is the most likely problem?

- **a.** Squirrel living under array
- **b.** **Improperly installed connectors**
- **c.** Cold temperatures below -40C
- **d.** Snow and rain

An arc fault in this case would most likely be caused by improperly installed connectors. If the connectors were not clicked into place, it could cause an arc fault. A squirrel would be a good second choice answer, but it may be difficult for a squirrel to unlock a properly installed connecter – what tool could it use? Cold temperatures will probably not cause this type of problem and also unlikely would be snow and rain.

33. Bad insulation around a wire with cracks would likely lead to?

- **a.** Voltage drop
- **b.** **Ground fault**
- **c.** Decrease in ampacity
- **d.** Increased insulation resistance

Cracking and problems with the insulation of the wire will likely lead to a decreased insulation resistance and ground faults. Ampacity and voltage drop are characteristics that are controlled by the copper or aluminum part of the wire, not the insulation.

34. Battery storage systems shall have, according to the NEC

- **a.** Label with weight of the batteries
- **b.** **Illumination not controlled by automatic means only**
- **c.** Goggles on site at all times
- **d.** An energy storage maintenance logbook on site at all times out together by a qualified person

480.10(G): illumination states that illumination shall be provided for working spaces containing battery systems. Illumination cannot be controlled by automatic means only. The word goggles is not in the NEC. We are not required to weigh the batteries, however, we should label the battery bank with the nominal voltage, maximum short circuit current, and date short circuit current calculation was performed.

An energy storage logbook and goggles would be good ideas, but are not required.

35. Which is 3-phase 240 delta with a high leg wire colors?

- **a.** Black, Red, Blue, White, Green
- **b.** Black, Red, White, Green
- c. **Black, Orange, Blue, White, Green**
- **d.** Brown, Orange, Yellow, Gray

Figure 9.1 Image 240 delta with high leg. *Source*: Author: wdwd. Licensed under the Creative Commons Attribution-Share Alike 3.0 Unported (https:// creativecommons.org/licenses/by-sa/3.0/deed.en), 2.5 Generic (https:// creative commons.org/licenses/by-sa/2.5/deed.en), 2.0 Generic (https:// creative commons. org/licenses/by-sa/2.0/deed.en) and 1.0 Generic (https://creativecommons. org/licenses/by-sa/1.0/deed.en)license.https://upload.wikimedia.org/ wikipedia/ commons/4/43/High_ leg_delta_transformer.svg

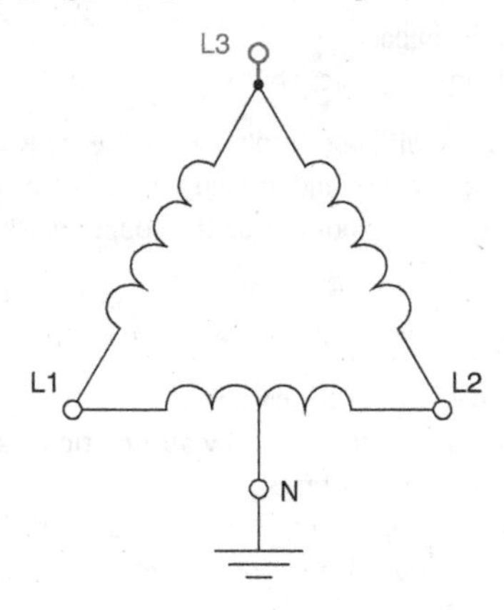

240 delta with a high leg is when you have 240 delta with one part of the triangle being like the 120/240 split phase that we bring into our house. The corner of the triangle opposite of the 240 side that is grounded in the middle is at a higher voltage than the other two phases that surround the grounded in the middle phase. When measuring voltages, line to line for all three phases, A, B, and C would all be 240. If we measure the high leg to neutral we would get 208V and if we measure one of the other phases to neutral we would get 120V. This high leg is often confused and called a stinger, since many electricians have thought it was 120V to neutral when they end up getting stung by 208V.

Figure 9.2 Image 240 delta with high leg at top and voltage. *Source:* Author: Gargoyle888. Licensed under the Creative Commons Attribution-Share Alike 3.0 Unported (https://creativecommons.org/ licenses/by-sa/3.0/deed.en) license. https://upload.wikimedia.org/ wikipedia/commons/e/ea/CenterTappedTransformer.svg

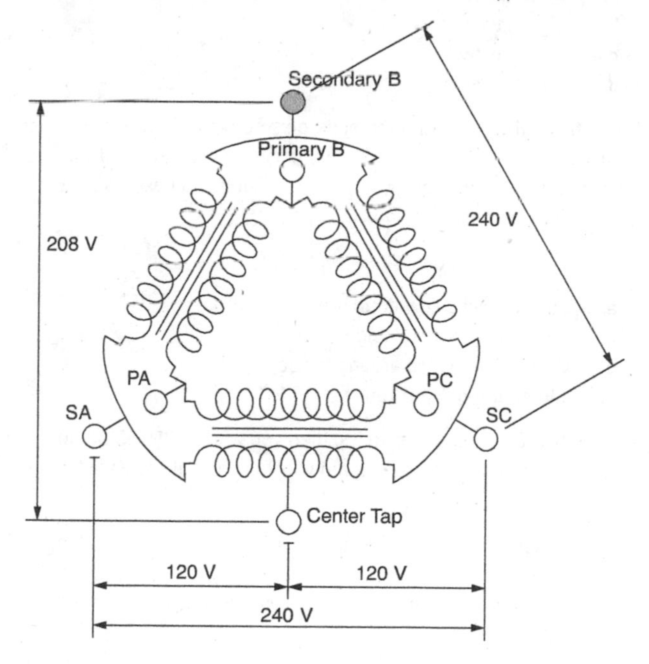

36. What are the usual wire colors for 277/480 3-phase power?

- **a.** Black, Red, Blue, White
- **b.** Black, Orange, Yellow, Gray
- **c.** **Brown, Orange, Yellow, Gray**
- **d.** Black, Orange, Yellow, Black

When you think of an orange wire, think that you could be working with something more dangerous. Orange is not only the stinger for 240 delta high leg, it is also present with 480V power.

Brown is line one, orange is line two, yellow is line three, and gray is the neutral. We are used to neutral being white, but it can be gray and in this case with 277/480, neutral is gray.

37. At what point would rigid PVC require expansion joints if expansion is over?

- **a.** **¼" or greater**
- **b.** ½" or greater
- **c.** ¾" or greater
- **d.** 1" or greater

To learn about different wiring methods, go to Chapter 3: wiring methods and materials and then find Article 352: rigid polyvinyl chloride conduit: type PVC. 354.44: expansion fittings tells us that with ¼" or greater expansion, we will be required to have expansion fittings.

38. Which type of electrode is unacceptable?

- **a.** Concrete encased electrode
- **b.** Steel frame of a building
- **c.** Copper wire buried around house
- **d.** **Aluminum ground rod**

Look to 250.52: grounding electrodes, then you will see 250.53(B): not permitted for use as grounding electrodes and you will see that there are three things listed:

1. Metal underground gas piping systems
2. Aluminum
3. The structures and structural reinforcing steel described in 680.26(B)(1) and (B)(2).

Article 680 like Article 690 is in Chapter 6: special qquipment and Article 680 only has to do with swimming pools, so we will be more concerned about **aluminum** here, which is, plain and simple, **not allowed as an electrode!**

39. What is the required width of working space in front of a PV system ac disconnecting means?

- **a.** The width of the ac disconnecting means or 30", whichever is less
- **b.** 30"
- **c. The width of the ac disconnect or 30", whichever is greater**
- **d.** The width of the equipment

110.26(A)(2): width of working space says:

> The width of the working space in front of the electrical equipment shall be the width of the equipment or 30 inches, whichever is greater. In all cases, the work space shall permit at least a 90 degree opening of equipment doors or hinged panels.

40. Which of the following is typically the best solution for making wires inaccessible on a ground mount?

- **a.** Security guards
- **b.** The neighbor's electric fence
- **c. Covering over the backs of modules**
- **d.** Alarm system

The exposed wiring between modules and along the backs of arrays should be inaccessible. For a ground mount, this usually means erecting a fence. Another acceptable method is covering the backs of the modules, so that the wires are inaccessible. Other methods include putting a lattice structure from the sides of the modules down to the ground. The neighbor's electric fence is not something you can depend on and neither are security guards and alarms.

41. What is the smallest allowable equipment grounding conductor going in conduit with 17 PV source circuits each with a short circuit current of 8A going in a single conduit to a combiner?

- **a. 14 AWG copper**
- **b.** 12 AWG copper
- **c.** 6 AWG copper
- **d.** 10 AWG copper

There are two places we can look to answer this question. First, we will go to 690 Part V: grounding and bonding, then we will look in Section 690.45: size of equipment grounding conductors. Here, it tells us that we should go to 250.122 and that the smallest that an equipment ground conductor can be is 14 AWG. Since the fuse size was not mentioned and we know that we need fuses when there are more than two strings in parallel at the dc combiner, then we will calculate the fuse size as 156% of Isc = 8A x 1.56 = 12.5A, which we will round-up to the next common fuse size of 15A. 15A is a typical fuse size for a combiner. We can see the common fuse sizes at Table 240.6(A). Table 240.6(A) is new and we would have the same information in Section 240.6 previously, but it is nicer to have a table.

Table 250.122: minimum size for equipment grounding conductors is a good table to remember where to look. Here we see that **with a 15A OCPD, we should have at least a 14 AWG equipment grounding conductor**.

The requirement we often see for 6 AWG copper equipment grounding conductors is for when the EGC is in free air and subject to damage, not when in conduit. We can have 6 AWG equipment grounding conductors going from the array to a junction box and then we can transition those 6 AWG conductors to 14 AWG. Inspectors and building departments do not always understand this and may have more stringent requirements.

42. What is the UL listing for inverters?

- **a.** 1703
- **b.** **1741**
- **c.** 2703
- **d.** 2741

UL 1741 is for inverters, UL 1741 SA is special for inverters in California that offers more grid support, UL 1703 is for PV, and UL 2703 is for racking systems.

43. In which situation is a bonding jumper not required when connecting EMT to a junction box in a system with over 250V?

- **a.** When the EMT is aluminum
- **b.** When using a WEEB
- **c.** When there are ringed knockouts
- **d.** **When there are no concentric knockouts**

Article 250: bonding and grounding contains Part V: bonding, which contains Section 250.97: bonding over 250V, which refers us to using the same methods as for services in Section 250.92(B): method of bonding at the service, except for (B)(1). 250.92 (B) says:

> **Bonding jumpers meeting the requirements of this article shall be used around impaired connections, such as reducing washers or oversized, concentric, or eccentric knockouts.**

Concentric definition: Multiple circles sharing a center.

44. What is NOT an example of when an interactive inverter would be operating off the MPPT point of the IV curve?

- **a.** Clipping
- **b.** Overheating
- **c.** **STC operation**
- **d.** Reducing power output for a non-exporting system

Any time an interactive inverter reduces power, it will have to work off the IV curve, unless it would be sending power somewhere else, so the best answer here is the situation where the inverter is an example of losing power, which is when it is operating at STC. Clipping is when you have more PV potential than the inverter can put out, so the inverter will have to work off the MPP. This is the same as when the inverter gets too hot, so it will not overheat and damage the inverter, the inverter will decrease power by working at a higher voltage and lower current than the MPP, thereby reducing heat in the inverter. Since current creates heat, the higher voltage will keep the inverter cooler. This is usually how an inverter will reduce power, in all of these cases is by moving to the right on the IV curve. Also when an inverter has to reduce power because the utility will not allow exporting, it will reduce power by moving to the right on the IV curve. On a question like this, beware of the word "NOT" and realize that three answers are things that will do what the question is asking.

45. Which is the best tool for checking for intermittent ground faults?

- **a.** Digital multimeter
- **b.** Clamp on ammeter
- **c.** **Megohmmeter**
- **d.** Two alligator clips and a jumper

A megohmmeter, otherwise known as a megger or an insulation tester, is best for checking for intermittent ground faults, because with an intermittent fault, the major fault is not usually there and the insulation tester can test the insulation. We would like millions of ohms of resistance in our insulation and, if we do not, then it can indicate that there is a problem or a potential problem.

46. PV system disconnects in different locations than the main service disconnect will need what?

- **a.** OSHA Certification
- **b.** **A plaque or directory**
- **c.** UL 1703 listing
- **d.** IEC label

If the disconnects are in different locations, they will need a plaque or directory, so that first responders and others will know that when they turn off the first switch, there are others to also turn off.

47. At what point must conductors have the ability to be rapidly shut down to less than 30V in less than 30 seconds outside of a building?

- **a.** At the module level
- **b.** 10 feet from the array
- **c.** 3 feet from the array
- **d.** **1 foot from the array**

690.12(B) states that the conductors must be controlled outside of the array boundary. The array boundary is also defined here as being 1 foot from the array. If inside a building, then the conductors must be controlled within 3 feet of the entrance to the building. In the 2014 NEC, the correct answer would have been b, 10 feet from the array. The 2019 requirements of the 2017 NEC require what we call module level shutdown, which is 80V in 30 seconds inside of the array.

48. At 33° latitude with a latitude −15° tilt and peak sun hours at that tilt being 4.5, what would be the minimum kW system size needed to produce 75% of the energy if you considered a derating factor of 77%? Assume that the required amount of energy to be used is 9000kWh per year.

- **a.** **5.3kW**
- **b.** 7.1kW
- **c.** 5.6kW
- **d.** 11.2kW

The numbers that we will use to answer this question are:

- 4.5 PSH
- 77% derating = 0.77
- 75% of energy to be made = 0.75
- 9000kWh per year to be used

The two things we are going to do are, first, to determine how much energy we need to make and, second, to determine how much energy 1 kW will make. Then, third, we will divide the energy we need to make into the energy 1 kW will make to get our answer.

First: 9000kWh x 0.75 = 6750kWh per year we need to make.
Second: 4.5PSH x 0.77 derating x 365 days = 1,265kWh/kWp/yr.
Third: 6750kWh/yr/1265kWh/kWp/yr = **5.34kW of PV is the answer.**

Look out for irrelevant information. We were given PSH at latitude -15°, but it does not matter for this calculation what the value is for latitude -15° in order to answer this question.

49. Battery room doors must?

- **a.** **Open in the direction of egress**
- **b.** Never be locked
- **c.** Always be locked
- **d.** Have ventilation built into the door

480.10: battery locations includes Section 480.10(E): egress, which states:

> A personnel door(s) intended for entrance to, and egress from, rooms designated as battery rooms **shall be open in the direction of egress** and shall be equipped with listed panic hardware.

Battery rooms that contain lead-acid batteries, which Article 480 evolved from, can be subject to explosion from hydrogen gas. These rooms containing lead-acid batteries require ventilation, however, there is no requirement that the ventilation has to be from the door.

50. Which is NOT required by the NEC for a PV array installation?

- **a.** Grounding of 2703 listed rail systems
- **b.** **Lightning protection**
- **c.** Torque wrench as required by manufacturers' instructions
- **d.** Module frames that are at ground voltage

Lightning protection is not covered in the NEC, it is covered in NFPA 780.

51. Given a 225A busbar and a 200A main breaker, what is the most current an inverter can have assuming a load-side, center-fed busbar?

- **a.** 32A
- b. **56A**
- c. 70A
- **d.** 25A

This is covered by the 120% rule 705.12(B)(2)(3)(b). When the 2014 NEC came out, we were not allowed to apply the 120% rule to center-fed busbars, however, mid-cycle there was a TIA, which is a Temporary Interim Amendment (which is really a permanent change to the NEC). This amendment allowed us to apply the 120% rule to center-fed busbars as long as we only put backfed breakers on one side of the busbar (and not both).

The rule states that 125% of the inverter current plus the main breaker cannot exceed 120% of the busbar rating.

Here is the math:

225A busbar x 1.2 = 270A
270A – 200A main = 70A allowance
70A / 1.25 = **56A inverter**

52. Given the following voltage measurements of a 3-phase 4 wire system, which is the high leg?

- **a.** Phase A to neutral 120V
- **b.** Phase A to phase C 240V
- **c.** Phase C to neutral 120V
- **d.** **Phase B to neutral 208V**

With a 3-phase 4 wire system you measure phase to neutral with every phase but one and get 120V, when you can get surprised is when you measure the phase to neutral that is 208V. That "high leg" is to be marked orange.

53. How do you achieve desired torque settings for connectors?

- **a.** Using an impact driver
- **b.** **Using a calibrated torque tool**
- **c.** Using a differential metered torque sensor
- **d.** Using a trained wrist

Using a calibrated torque tool is the correct answer, as to manufacturers' installation instructions.

54. You would like to produce 80% of the energy at a house with an annual consumption of 7500kWh per year. The slope of the roof is 5:12 and is facing 180 azimuth at latitude 38. With 82% system efficiency, what would be the required amount of PV? (Use the following information.)

Solar insolation and tilt for location at latitude 38
Tilt 0 = 4.9 PSH
Latitude – 15 = 5.2 PSH
Latitude = 5.4 PSH
Latitude + 15 = 5.1 PSH

- **a.** **3.9kW**
- **b.** 4.1kW
- **c.** 3.1kW
- **d.** 5.0kW

In order to convert roof slope, we can take a good guess or use a trig function, which is easy. Here is what you punch into your calculator to get the degree angle of a 5:12 slope roof:

$$5/12 = 0.41667$$
$$0.41667 \text{ inverse tan} = 22.6° \text{ slope}$$

So 15° greater than 22.6° is about 38°, so we will use the information for latitude –15°, which is 5.2 PSH.

$$5.2 \text{ PSH} \times 82\% \text{ efficiency} = 4.26 \text{ ac kWh per day}$$
$$4.26\text{kWh} \times 365 \text{ days} = 1555\text{kWh/kWp/yr}$$

We want to produce 80% of 7500kWh = 6000kWh:

6000kWh/1555kWh/kWp/yr = 3.86kW

55. What type of inverter/array grounding system includes fuse grounded inverters and most inverters without transformers?

- **a. Functional grounded system**
- **b.** Ungrounded system
- **c.** Isolated grounding system
- **d.** Non-isolated grounding system

See 690.47(A): informational note. Most PV systems installed in the past decade are actually functional grounded systems rather than solidly grounded systems as defined in this Code.

Therefore, a fuse grounded inverter and modern non-isolated inverter without a transformer are both part of functional grounded PV systems. A non-isolated inverter is the modern transformerless inverter, however, non-isolated inverters do not include fuse grounded inverter systems.

56. When troubleshooting four strings, you notice that three strings have slightly less current than the fourth string, but that when you measure voltage,.the fourth string has 0V and the other three strings have normally expected voltage. What is the likely problem?

- **a.** Ground fault on the fourth string
- **b. Short circuit on the fourth string**
- **c.** Fourth string disconnected
- **d.** Forth string open circuited

If there is a short circuit with four strings going to a combiner, then likely the fuse will be blown and the shorted string would have a slightly higher voltage than the other strings, since Isc is higher than Imp. Also, when voltage is measured, the shorted string would measure 0V.

57. Using standard crystalline modules with 6" solar cells (156mm), what would be the most number of modules that you could parallel to a charge controller without PV source circuit fusing?

- **a.** 0
- **b.** 3

c. 1
d. 2

In most all cases, you can only have 2 PV source circuits combined together without needing fuses. In this example, since it is a 12V system using 12V modules, the PV source circuits consist of a single module each (just like a PV source circuit with a typical microinverter or dc-to-dc converter). In rare cases, some thin film PV modules can have more than two PV source circuits in parallel.

Most solar PV installed in the world is with 6" crystalline silicon solar cells, having a maximum series fuse rating that is 156% of Isc and then rounded-up to the next common fuse size, which is usually 15A or 20A. With these factors, we will always have two in series not requiring fuses when combined and three in series will need fusing when combined.

58. If a thin film PV module has a short circuit current of 1.8A and a maximum series fuse rating of 15A, what would be the maximum number of PV source circuits combined at a dc combiner without needing fusing?

a. 2
b. 8
c. 1
d. 3

690.9: overcurrent protection, brings us to 690.9(A): circuits and equipment, which says we need overcurrent protection unless the circuits have sufficient ampacity for the highest available current, such as when connected to current limited supplies (like a PV module).

690.9(A): exception (2) states that:

> **The short-circuit currents from all sources do not exceed the ampacity of the conductors and the maximum overcurrent protective device size rating specified for the PV module.**

Since the PV module has an unusual 1.8A Isc and the PV module maximum series fuse rating is 15A, then the calculation is as follows:

15A max series fuse rating/1.8A Isc = 8.3 in parallel

So, the maximum in parallel in this case could not be more than eight in parallel. We would always round-down for this calculation.

59. Array operating current is 80% of expected on a PV output circuit. Using a clamp ammeter, all of the PV source circuits measure the same current except for one string, which measures 0A. The installer then pulls all of the fuses and does continuity tests on all of the fuses and they all check out good. What is the most likely scenario?

- **a.** Shorted bypass diode
- **b.** Module wired with reverse polarity
- c. **Disconnected wire**
- **d.** Intermittently blown fuse

Since there are 0A on a string, it is probable that there is **no continuity within a string** (PV source circuit). Since the fuses were tested for continuity, it is unlikely that a fuse would work intermittently. **The most likely scenario here is a disconnected wire.**

Modules can still work with reverse polarity and, back in the day, before we had MC connectors coming out of most modules, it was not completely uncommon for someone to wire a module with reverse polarity. What happens when a module is wired in reverse polarity is that it will subtract voltage from the string rather than add, so for example with four 40V modules in a string, you should get 160V, but if one of the modules were subtracting voltage instead of adding voltage, you would get 80A. That was once a lab test that we did in class with modules that were connected with terminals, rather than MC connectors.

60. Two PV source circuits of 10 220W 60 cell PV modules in series are connected to a single MPP inverter that was installed in 2008. You were asked to check on production and noticed that the array was shaded 1 foot from the edge on the short edge of most of one of the strings at noon in the winter. The modules were mounted in portrait. If you measured 1kW of production in this scenario, about what would be the production without the shaded strip running down the short edge?

- **a.** 1kW
- **b.** 1.5kW
- c. **2kW**
- **d.** Greater than 2.5kW

Most modules when shaded along the short edge in this type of situation will have the bypass diodes activate and will not have current from the shaded string, even if the entire module is not shaded. The best answer here is 2kW.

61. A rooftop PV array is to be installed in a track home, where the HOA requires that no conduit be visible from outside of the building. Which of the following wiring methods will be permitted to run conduit in attic?

- a. **USE-2/RHW-2 in free air to junction box and then USE-2/RHW-2 in EMT**
- b. USE-2 in free air to junction box under modules and then USE-2 in EMT
- c. USE-2 in free air to junction box under modules then THWN-2 in rigid NMC
- d. USE-2/RHW-2 in free air to junction box under modules then THWN-2 in rigid NMC

USE-2 and many other conductors can be dual listed, which means it can act as two different wires. For instance, a common dual listing is THWN/THHN. **USE-2 alone is not allowed in conduit**. You can use **USE-2/RHW-2 for the entire run of the PV source circuit**. This is not commonly done, since USE-2/RHW-2 is more expensive and has a larger diameter than TJHWN-2, which is more commonly used. The reason that the THWN-2 answers do not work in this situation is because when PV source circuits are inside buildings, they need to be inside of a metal raceway, such as EMT, or in MC cable. Rigid NMC is plastic.

62. What is the main reason for the inverter bypass switch in a utility-interactive, battery-backup system?

- a. So that backed-up loads can operate during power outage
- b. To charge batteries with the utility
- c. **To allow backed-up loads to be used during inverter failure or maintenance**
- d. So that anti-islanding can be controlled

With a grid-tie battery backup system, the backed-up loads go through the multimodal inverter when the grid is operating during normal operation. If there were a problem with the inverter, it is necessary to flip a switch to bypass the inverter, so that the backed-up loads will get grid power.

63. 12 AWG USE-2/RHW-2 is lying on a tile roof and going 20 feet across the roof. What can be done to make this comply with the NEC?

- a. **Run the cable in EMT.**
- b. Install in Liquidtight Flexible NMC and secure every 6 feet.
- c. Support the cable on purlins.
- d. Put the cable under the tiles but above the roofing felt.

The best solution is to run the conductor in EMT. There could be an argument for the cable tray, but leaving the conductors exposed would be not recommended.

356.30 states that LFNC (Liquidtight Flexible Nonmetallic Conduit, when longer than 6 feet, must be secured every 3 feet. EMT should be secured every 10 feet.

Putting the wire on purlins or underneath the tile would likely be interpreted as a code violation, since it is not in a metal raceway, among other things.

64. On a partly cloudy day an inverter intermittently shuts off for periods of 5 or 10 minutes at a time? What is the most likely cause of this problem?

- **a.** PV source circuit too far from inverter
- **b.** Undersized dc conductors
- **c.** **Undersized ac conductors**
- **d.** Intermittent ground fault

If an inverter does not sense proper frequency and voltage from the grid, it will shut off. If the ac wire going to the inverter, then Ohm's Law will cause the inverter to shut off.

$$V = IR$$

The resistance is a product of the wire and when the wire has too much resistance, by being undersized, it creates more voltage drop. This voltage drop will increase when there is more current and having intermittent clouds will cause times of higher and lower current, causing higher and lower voltage drop. Voltage drop will cause the inverter to work at a higher voltage in order to overcome the resistance of the wire. If the voltage at the inverter is 10% higher than grid nominal voltage, then the inverter will shutoff and it will take 5 minutes for the inverter to come on again. Bright light can cause an increase in voltage drop, so if the light is still bright at the end of 5 minutes, the inverter may try and come on for an instant and then start the 5 minute clock over again, causing a 10 minute break for the inverter.

65. What would be the largest single phase inverter that you could put on a breaker in a 208V 125A subpanel with a 100A feeder breaker?

- **a.** 12kW
- **b.** 10.4kW
- **c.** 9600W
- **d.** **8320W**

This is a simple 120% rule question, with the trick here being that most people are used to doing this equation with a 240V system and, in this case, we have some new numbers because the voltage is 208V.

Here is the math:

125A panelboard x 1.2 = 150A
150A – 100A = 50A
50A = 125% of inverter current
50A/1.25 = 40 = 50A x 0.8
40A x 208V = 8320W

66. When connecting flexible, fine-stranded copper wire to an inverter, combiner box, or battery terminal lug, the installer must ensure that the terminal lug is rated for?

a. Dc use only
b. 75C
c. **Fine-stranded wire**
d. 90C

Flexible fine-stranded wire when attached to improper terminals will loosen up over a short period of time and never make a good connection. Often with flexible fine-stranded wire compression connectors are used. It is very important to use equipment that is rated for the correct task.

67. An interactive residential PV system is installed in Alaska on a single family home. The system is commissioned in July and the open circuit voltage measures 580V. If there is a utility failure in the winter, what could happen?

a. The system would deliver more than normal power.
b. **The inverter warranty would likely be voided.**
c. The inverter would be likely to indicate a ground fault.
d. The inverter would limit current.

PV systems on single family homes are limited to 600V. In a place with cold winters, the voltage can be substantially more in the winter, even greater than 20% more than the voltage measured on a summer day. In this scenario, if the open circuit voltage measured 580V in the summer, it could easily measure well over 600V in the winter if the utility went down. When the utility goes down, the PV system goes to open circuit voltage, which is substantially higher than operating voltage. If the voltage on the dc bus of an inverter that is turned off, due to a utility outage will void the warranty of the inverter and also potentially damage the inverter and become a danger.

68. If you install an 8 AWG grounding electrode conductor, which of the following is the best method?

- **a.** In free air when not exposed to physical damage
- **b.** **In RMC, IMC, PVC, RTRC-XW, EMT, or cable armor**
- **c.** In USE-2 cable
- **d.** Securely fastened to the surface of the building

250.64: grounding electrode installation contains 250.64(B)(3) which tells us: **"Grounding electrode conductors smaller than 6 AWG shall be protected in RMC, IMC, PVC, RTRC-XW, or cable armor."**

Also, we can see that 260.64(B)(1): not exposed to physical damage tells us that 6 AWG or larger **not** exposed to physical damage may be run along the surface of the building without metal covering or protection.

Additionally 260.64(B)(2): exposed to physical damage tells us that 6 AWG or larger exposed to physical damage shall be protected in RMC, IMC, PVC, RTRC-XW, EMT, or cable armor.

As we can see here, the rules are different for grounding electrode conductors and equipment grounding conductors, they are more restrictive and careful with grounding electrode conductors, requiring 6 AWG and larger to be protected when exposed to physical damage.

69. What is the primary purpose of rapid shutdown for PV systems?

- **a.** Preventing ground faults
- **b.** Preventing firesnlsl
- **c.** **First responder safety**
- **d.** Preventing shocks for solar installers, electricians, roofers, and other construction workers

The primary purpose for rapid shutdown of PV systems is first responder (firefighter) safety.

70. What is the maximum circuit current for a dc-to-dc converter source circuit?

a. **Manufacturers' continuous output current rating**
b. Isc
c. 125% of Isc
d. 156% of Isc

690.9: circuit sizing and current brings us to 690.9(A): calculation of maximum circuit current, which contains 690.8(A)(5): DC-to-DC converter source circuit current, which tells us that the maximum current shall be the dc-to-dc converter continuous output current rating. Using Isc in a calculation is for PV, not for dc-to-dc converters (unless someone invents something in the future).

Practice exam 2

Chapter 10

70 questions, four hours

This practice exams in this book are by far the most valuable part of this book for those of you wanting to be NABCEP Certified. Once you get the basics down, the most efficient way to study for an exam is by taking practice exams and going over the answers in detail in the following chapter. Much of the basic concepts that we use are first introduced in the *Solar Photovoltaic Basics* book. Here we are putting those basic concepts in an advanced, realistic context. This book contains two full-length 70-question practice exams and a bonus exam, followed by detailed answers and explanation.

Set your timer for 4 hours. Sharpen your pencil. Use your NEC Codebook.

Ready, set, go!

1. You are installing a concrete encased electrode (ufer) on an inverter pad. There is a conductor that is coming out of the concrete to make a connection to attach the electrode to the PV system. What is the largest that this conductor has to be if it is copper?

- **a.** 4 AWG
- **b.** 3/0
- **c.** 8 AWG
- **d.** 6 AWG

2. You are building a new house in the woods with a ground-mounted PV system 1000 feet away in an opening. You decide you would like to use a ground ring, since you have a lot of 2 AWG wire left over. How deep must you bury the ground ring?

- **a.** 18" in conduit
- **b.** 30" or more

c. 8 feet or more
d. As deep as the foundation

3. The inverter went over voltage in Kalamazoo one cold winter due to global warming. The warranty was voided and the installer was responsible for buying the new inverter. The maximum input voltage of the inverter was 1000V. The PV module specifications at STC were: power = 245W, Vmp = 30.1V, Imp = 8.2A, Voc = 37.7V, Isc = 8.7A. The temperature coefficient for the monocrystalline silicon module was –0.121V/C. There were 24 modules per PV source circuit. How cold did it get when the inverter got to be 1000V dc?

   a. –8C
   b. 0C
   c. –18C
   d. –12C

4. Four 240V 3kW utility interactive inverters are connected to a 100A subpanel on a house that is fed by a 100A breaker. The subpanel was installed for solar and there are no loads on it. What must the label on the subpanel say?

   a. WARNING! INVERTER OUTPUT CONNECTION; DO NOT RELOCATE THIS OVERCURRENT DEVICE
   b. WARNING! DO NOT CONNECT MULTIWIRE BRANCH CIRCUITS!
   c. WARNING! ELECTRIC SHOCK HAZARD DO NOT TOUCH TERMINALS TERMINALS ON BOTH THE LINE- AND LOAD-SIDE MAY BE ENERGIZED IN THE OPEN POSITION
   d. WARNING! THIS EQUIPMENT FED BY MULTIPLE SOURCES. TOTAL RATING OF ALL OVERCURRENT DEVICES, EXCLUDING MAIN SUPPLY OVERCURRENT DEVICE, SHALL NOT EXCEED AMPACITY OF BUSBAR

5. A monitoring unit is positively grounded. What would be the appropriate color coding and where do the disconnects open the circuit?

   a. Grounding conductor should be white and positive and negative should be disconnected by a single switch. Overcurrent protection on negative and positive.
   b. Disconnects on negative. The negative can be black. The grounded conductor is white. Overcurrent protection on negative.

c. Blue negative, white positive, yellow ground with green stripe, and disconnect positive and negative. Overcurrent protection on negative.
d. White negative, black positive, green ground, and disconnect opens black and white conductors. Overcurrent protection on positive and negative.

6. An interactive UL 1741 listed inverter is turning on and off on sunny days during what would be expected to be the time of day for peak performance. You are sent to the jobsite to troubleshoot the problem. Of the following, which would be the most likely scenario?

a. Voltage too low at the ac terminals of the inverter
b. Voltage too high at the ac terminals
c. Voltage too high at the dc terminals
d. Frequency too high at the terminals

7. Which is the best way to determine the state of charge of a typical flooded lead-acid battery:

a. Hydrometer testing
b. Test voltage at minimum load
c. Charge battery for 25 minutes and then immediately measure voltage with a digital multimeter
d. Run down battery 50% then test depth of discharge and calculate 1 – depth of discharge = state of charge

8. You are assigned to install a PV system with a combiner box at a car wash in an outdoor area subject to high-pressure water, air blown dust, and high temperatures. Which type of enclosure should you specify at the electrical supply house?

a. NEMA 3R
b. NEMA 4X
c. NEMA 11
d. UL 1703

9. You are commissioning a PV system in a humid location near salt water and you notice that in a positive grounded PV system, when the system is on the grounded conductor measures near 0V to ground and the negative wire measures 412V to ground. When you turn the system off at the dc

disconnect, the grounded conductor at the combiner box stays near 0V and the negative terminal at the combiner box goes up to 494V. What is the problem with this situation?

a. Negative ground fault at array
b. Positive ground fault at combiner box
c. System ground has come loose and is shorting intermittently
d. System may be within normal limits

10. There is a PV system that is not working and you are sent out to troubleshoot the problem. The inverter is about 4 years old and is the type with a ground-fault detection and interruption fuse (GFDI). The inverter is a grounded inverter. The light is blinking that indicates a ground fault is present. You are about to go on the roof and you just realized that you do not have fall protection with you. Also, you notice that the dc equipment grounding conductor was removed from the inverter by one of your coworkers that tried to troubleshoot the system the day before. What might be a hazard with this system?

a. The array frames and racks may be energized with reference to ground.
b. Since the grounded conductor cannot be disconnected, at least you know that the white wire is at the same potential as the equipment.
c. It is likely that the array voltage is 0 on positive and negative grounded because of a short circuit.
d. The green and the white wire will have the same voltage.

11. You are required to put posts into concrete and the specifications call for holes that are 18" in diameter and concrete going 5 feet deep. The holes are going to have 3" diameter pipe going 5 feet down. If you are going to have 188 of these posts, then how many cubic yards of concrete do you need?

a. 1617 yards
b. 64 yards
c. 60 yards
d. 102 yards

12. For a large ground-mount system every eight modules will need two ballast blocks. Each ballast block is made from concrete that weighs 140lbs per cubic foot. The dimensions of each block are 2 feet 3" north to south, 3 feet 2" east to west. If the ballasts for the section of eight modules have to weigh 3824lbs, to keep the array stable, then how deep must we pour the concrete into the forms?

- **a.** 23"
- **b.** 46"
- **c.** 72"
- **d.** 48"

13. On a ballasted ground-mounted PV system, the width of the footing would be wider due to

- **a.** Soil load bearing
- **b.** Frost line
- **c.** Wind uplift
- **d.** Insulation

14. Fidelity Roof Company is a third generation roofing company with a reputation for quality. The roof being looked at for solar has been well taken care of by three different generations of roofers and each generation has put another roof on the building. What would be the main concern to look for when determining if solar should go onto this composition asphalt shingle roof?

- **a.** The buildup of roofing materials is too thick for a good attachment.
- **b.** Redwood lumber will get too weak over the span of 60 years.
- **c.** Too much weight from layers of roofing.
- **d.** There are no good waterproofing materials compatible with asphalt shingles.

15. The owner of a house complains of low performance. You come to the house and try to figure out what the problem is. What would be the best thing to do first of the following?

- **a.** Test the current of each string
- **b.** Test the voltage at the inverter
- **c.** Make sure connectors under modules are secured
- **d.** Look at the grounding

16. A giant 100 feet tall American flagpole with a giant flag waving side to side is directly south of a solar array at 38° latitude. At solar noon on winter solstice, the solar elevation angle is 28°. You are going to put solar on a rooftop and would like to make sure that the flagpole does not shade the array. Your rooftop is 65 feet high and the roof is a relatively flat roof. How far away from the flagpole should the solar array be located?

a. 45 feet
b. 66 feet
c. 73 feet
d. 10 yards

17. You ran across a free 10kW inverter and you only have 3kW of PV to install. Since the inverter was free, you decided to install it on your house. Your rooftop can barely fit 3kW on it and there is no way you will ever put more PV on that inverter. What is the size of the breaker you will use for that system? Assume that the inverter is 240Vac.

a. 15A
b. 20A
c. 30A
d. 60A

18. You are asked to connect on the supply side of a 2000A breaker. What should you do in order to connect this system?

a. Pull the meter while wearing an arc-flash suit
b. Disconnect the current transducers
c. Open utility disconnects
d. Wear PFAS and steel-toed boots

19. If you had an off-grid system with a 10kWh battery bank, a ground rod, an ac coupled inverter system, an inverter that is connected to the inverter with a 300Adc disconnect, and a 250A OCPD between the inverter and the battery bank, what would be the correct size for a copper equipment grounding conductor between the battery bank and the inverter?

a. 4AWG
b. 6AWG
c. 250kcmil
d. 1/0AWG

20. You are working on a jobsite and the wrong inverter was sent, because the old inverter was thought to be less safe and less efficient by the business owner. What must you do to make sure that the new ungrounded inverter is installed correctly when adapting the plan from the grounded inverter?

- **a.** Switch to a larger grounding electrode conductor.
- **b.** Not use 90°C rated USE-2 for your source circuits on the roof outside of conduit.
- **c.** Tap the transformer on the ungrounded inverter.
- **d.** Switch the white wire from the negative to the positive.

21. A 1MW inverter with 1200kW of PV is operating on a sunny day with 780W per square meter of irradiance and a cell temperature of 45°C and dc to ac derating of 95%. What is the expected inverter output power? Assume a typical temperature coefficient of power of -0.45%/C

- **a.** 1MW
- **b.** 1.1MW
- **c.** 908kW
- **d.** 809kW

22. You are studying notes from a site survey and note that the panelboard busbar is rated at 425A and the main breaker is rated for 400A. The service voltage is 240Vac. What would be the size of the largest ungrounded inverter that you could connect on the load-side?

- **a.** 110kW
- **b.** 21kW
- **c.** 33kW
- **d.** 26.4kW

23. When using a safety monitoring system, a designated safety person must

- **a.** Be in charge of installing a line with flags a minimum of 6 feet apart on a line.
- **b.** Inspect fall protection systems and run weekly toolbox meetings when there are groups of 11 employees or more.
- **c.** Have no other duties, be on the same level of work, and close enough for oral communication.
- **d.** Be present for all work over 6 feet in height.

24. You are called to a jobsite and troubleshooting an inverter that is not sending power back to the grid. You were told by a coworker that inside of a dc disconnect that the white wire to ground is 402V and the black wire to ground is less than 1V. What is the most likely problem?

- **a.** The inverter is turned off and just needs to be turned on by closing the dc disconnect.
- **b.** There is a ground fault on the grounded conductor.
- **c.** There is a ground fault on the ungrounded conductor.
- **d.** There is a positive to negative short circuit.

25. A broken grounded inverter has been replaced with a newer ungrounded inverter. You were not the one to replace the inverter. Six months after the inverter has been replaced, the monitoring system has determined that the system suddenly has stopped producing energy. Your supervisor has checked everything and said he cannot find the problem. He said that voltage to ground on the white dc wire was 0V. What else would you expect to find and what would you expect the problem to be?

- **a.** There is a ground fault and the equipment would be expected to be the same voltage as the white wire in this scenario.
- **b.** There is a short and the ungrounded conductor will be at the same voltage as ground until the problem is fixed.
- **c.** There is a short and the ungrounded conductor will be at the same voltage as the inverter cabinet until the problem is fixed.
- **d.** There is a ground fault on the ungrounded conductor and the white wire is therefore ungrounded by the fault.

26. What is the safest way to connect load- and line-sides of ac and dc disconnects on a utility interactive system?

- **a.** Load-side of dc disconnect to inverter and line-side of ac disconnect to inverter
- **b.** Line-side of dc disconnect to inverter and load-side of ac disconnect to inverter
- **c.** Load-side of dc disconnect to inverter and load-side of ac disconnect to inverter
- **d.** Line-side of dc disconnect to inverter and line-side of ac disconnect to inverter

27. There is a building that you would like to put different inverters on for different tenants. What would be required in this situation?

a. There must be six or fewer PV disconnects to turn off all of the PV systems and all of the systems must be in the same enclosure.
b. There must be 12 or fewer PV disconnects.
c. There shall be a maximum of six disconnects to turn off all of the PV systems and the disconnects need to be in the same location.
d. There may be an unlimited number of disconnects as long as they are the anti-islanding type.

28. A PV dc disconnect may go in any room in the house except for the

a. Kitchen
b. Bathroom
c. Bedroom
d. Nursery

29. What are the rapid shutdown voltage limits for inside the array boundary and outside the array boundary?

a. 30V in 30 sec. inside and outside the array
b. 30V inside array and 80V outside of the array
c. 80V inside the array and 30V outside of the array
d. 80V in 30 sec. inside and outside of the array

30. You are using 270W microinverters and 300W solar modules on a residential service at 240Vac. You want to make sure that you are using the right number of microinverters for the cable that the microinverters are connected to by using the NEC. If the cable is capable of handling 25A, then how many microinverters can you have on the cable?

a. 25
b. 17
c. 18
d. 22

31. You are installing a PV system on a building and the plans call for AC cable in a wet location. What is the best solution?

a. Use AC cable with raintight connectors.
b. Change to EMT.

c. Use AC cable with weathertight connectors.
d. Use PV wire inside of AC cable.

32. What is the smallest size of EMT that you can use for 12 current-carrying 12AWG THWN-2 conductors and one 12 AWG equipment-grounding conductor?

a. 1"
b. 3/4"
c. 1.5"
d. 1.25"

33. There is a new LEED platinum building going up and there is a 1000A three-phase service that you are asked to connect solar to. There are multiple 200A subpanels in the building. What would be the most likely way you would connect a 5kW system to this building?

a. Load-side connection on a subpanel
b. Supply-side connection
c. Load-side connection on main panel
d. Feeder tap

34. With a 120Vac inverter on a house with a standalone PV system, can you use a 120/240V service panel and, if so, what must you do?

a. Yes, as long as you do not connect Line 1 to Line 2.
b. Yes, as long as you do not use multiwire branch circuits.
c. Yes, if you design the system to not overload the neutral.
d. No.

35. You are supervising a utility-interactive job using grounded inverters and there are two different electricians installing dc disconnects. The first electrician is installing the disconnects with the green ground screw that comes with the disconnect, connecting the grounded conductor bus to ground and the other electrician was not using the green ground screw. Which electrician was right and why?

a. The first electrician because, according to 110.3(B), equipment should be used according to directions.
b. The first electrician because the grounded conductor needs to be grounded.

c. The second electrician because using the screw would cause a ground fault.
d. The second electrician because it is a grounded inverter.

36. What is the best type of wire to use for series connections in a battery bank?

a. DLO
b. Heavy duty arc welding cable
c. THWN
d. Bare copper

37. Which of the following would be the least appropriate use of flexible fine-stranded cables?

a. Using with screw terminal lugs
b. Using with batteries
c. Using with compression connectors
d. Using with 4/0 wire

38. Which would be a wiring method appropriate for on a roof in sunlight and in an attic for PV source circuits?

a. Rigid non-metallic conduit
b. PVC
c. PV wire
d. EMT

39. The UL label on an inverter says that the inverter can be used for ungrounded arrays. In which situation can it also be used for grounded arrays?

a. If it will operate ungrounded, it can be used grounded, which is safer.
b. If the manufacturers' instructions say it can be used for bipolar arrays.
c. If the UL label also said it can be used for grounded arrays.
d. If GFCI is installed.

40. The dimensions of a 250W PV module are 1670 mm × 990 mm and there is a 5kW system. The wind uplift forces are 23lbs per square foot. The system is mounted in four rows with eight standoffs per row supporting the array. What would be the amount of uplift force per standoff?

- **a.** 120lbs
- **b.** 232lbs
- **c.** 256lbs
- **d.** 44 newtons

For questions 41 through 45, use the following manufacturers data:

Table 10.1 PV module data

Voc	37.8V
Vmp	31.1V
Isc	8.28A
Imp	8.05A
Temperature coefficient Voc	-0.30%/C
Temperature coefficient Vmp	-0.45%/C

Table 10.2 Inverter data

Power	5kW
Max input voltage	600V
MPPT voltage range	280–480V
Number of MPPT inputs	1

Table 10.3 Temperature data

Low design temperature	-13C
High cell design temperature	69C

41. What is the maximum number of modules in series?

- **a.** 14
- **b.** 12
- **c.** 13
- **d.** 22

42. What would be the minimum number of modules that you would put in series, so that the inverter works efficiently when it is hot?

a. 7
b. 12
c. 8
d. 10

43. If 20 modules will fit on the south-facing roof and 12 would fit on the southeast-facing roof, what would be the maximum amount of PV that you can use on this job with one inverter and these modules?

a. 3.8kW
b. 4kW
c. 5kW
d. 6kW

44. In the system design in Question 43 on the label at the dc disconnect, what number will we put for maximum system voltage?

a. 600V
b. 480V
c. 505V
d. 589V

45. For the system in Question 43, what is the smallest allowable PV source circuit conductor size for PV wire jumpers behind the array? Assume 33C ASHRAE hot design temperature, 75C screw terminals, and 15A fuses in the combiner. Assume copper conductors.

a. 10 AWG
b. 12 AWG
c. 14 AWG
d. 16 AWG

46. There is a 33 module 10kW ground mounted system with three rows of crystalline PV mounted in portrait. The modules are 72 cells with cells in the typical 6 × 12 arrangement. On a sunny December 18 at 2p.m., one out of three rows of modules is having one row of cells shaded. What would be the expected power loss from the shading assuming that all of the modules are connected to a single inverter with a single MPPT?

a. 4% loss
b. 33% loss
c. 50% loss
d. 90% loss

47. In a grid-tied system with battery backup, which of the following would be the most important attribute?

a. Having a charge controller that will power diversion loads when the system is in interactive mode.
b. Having a system that will equalize sealed lead-acid batteries.
c. Using a multimodal charge controller with an automatic transfer switch.
d. Having an inverter that will power all of the backed up load.

48. Which of the following would be the most important feature to have with a supply-side connection?

a. Have a service-rated fusible disconnect.
b. Do not exceed the rating of the main breaker.
c. Do not exceed 120% of the ampacity of the service conductors.
d. Do not exceed the rating of the busbar.

49. On a low slope roof ballasted PV system, there are rooftop combiners that can take up to 20 strings each. If the PV Isc is 8.25A and the Imp is 7.9A and the PV source circuits each have 12 modules in series, then how many strings can each combiner have if there is a fused disconnect on the PV output circuit that is 200A?

a. 15
b. 24
c. 20
d. 17

50. You are installing PV in a new laboratory for quantum computers where the temperature in the room is kept as cold as that in a freezer. You have conduit running through the cold room. What must you do for this rigid metal conduit that you would not otherwise have to do in a warmer environment?

a. Conduit fill calculations due to the absence of heat.
b. Fill raceway with material to prevent circulation.
c. Install a heating element into the conduit.
d. Allow air to circulate through the conduit.

51. You are installing residential solar systems in Phoenix, Arizona, and need to make sure that you do not put too few modules in series. The data from the PV module are: power 225W, Voc 48.5, Vmpp 41, Impp 5.5, Isc 5.9, temperature coefficient Voc –133 mV/K, coefficient Vmpp –156 mV/K. The inverter specifications are: maximum voltage 600V and MPPT operating voltage 240V to 480V. The low design temperature is –13C, the high ambient temperature is 44C and the high design cell temperature is 75C. What is the least amount of modules that you can put in series without sacrificing performance?

a. 11
b. 8
c. 5
d. 7

52. There are 15 PV source circuits and an equipment grounding conductor on a rooftop in sunlight in a circular raceway, which is elevated 3" above the roof going from the array to a combiner 22 feet away in a location with a constant breeze keeping the conduit cool. The ASHRAE 2% average high design temperature is 35C. The PV source circuits are 12 AWG THWN-2 and the ground wire is 12 AWG THWN-2. The terminals in the combiner box are rated for 75C. What is the maximum Isc for modules used in this system?

a. 7.65A
b. 10.2A
c. 16A
d. 24A

53. You are running conduit from an array to a transition box in the roof and there are no terminal temperature limits given on the terminals, the transition box, or the installation manual for the transition box. At the transition box, the conductors will change from PV wire to THWN as the circuits go from the roof to the attic. Should you consider terminal temperature limits or what should you use for terminal temperature limits?

a. Do not consider terminal temperature limits.
b. Use 60C terminal temperature limits.
c. Use 75C terminal temperature limits.
d. Use 90C terminal temperature limits.

54. What is the smallest copper equipment grounding conductor that can be used when running 10 source circuits in conduit with 10AWG THWN-2 wire 100 feet across a roof to a combiner box on the roof? The fuses in the combiner box are 15A and the combiner box terminals are rated for 75C.

a. 14 AWG
b. 10 AWG
c. 12 AWG
d. 6 AWG

55. A 10kW system with 250W modules in PV source circuits of 10 in series is producing about 75% of what was expected and you are sent to trouble-shoot the system. First, you turn off the system and check each string individually at the combiner box, which is at the inverter. One of the strings measures 0V and all the other strings measure within 2V of 351V. You then turn the system on and check the current of each string and notice that the string that measured 0V has more current than the other strings. All the modules are facing southwest at a tilt angle of 18°. What is the most likely problem?

a. Ground fault
b. Short circuit
c. Uneven number of modules in different PV source circuits
d. Polarity of one PV source circuit is reversed

56. What is the correct OSHA practice?

a. Have a safety meeting every day.
b. Tell workers about hazards.
c. Tell workers about their health and hospital plan.
d. Provide earplugs to all employees.

57. Who removes the lockout tagout (LOTO) tags?

a. The supervisor
b. A specially designated person
c. A licensed electrician
d. Whoever put the tags there

58. What is the biggest danger from seeing an arc from arm's distance?

a. X-rays
b. UV rays
c. Infrared rays
d. Shrapnel

59. Of the following, what is the safest thing to do first when checking wires on a roof after a ground fault has been indicated?

a. Remove fuses from combiner
b. Turn on inverter
c. Inspect wires on the roof
d. Measure voltage

60. What is the safest test of the following?

a. IV curve testing
b. Measuring current through the meter
c. Measuring current with a clamp amp meter
d. Measuring voltage

61. What is the device that is used for testing insulation of a conductor?

a. Megohmmeter
b. Refractometer
c. Pyrometer
d. Pyrheliometer

62. If the dc disconnect is not near the main service disconnect at the service entrance, what should be done according to the NEC?

a. Install another dc disconnect at the service entrance. Under no circumstances can the dc disconnect not be at the main service disconnect at the service entrance.

b. Include the location of all disconnects in the as-built plans that you will submit to the building department.

c. Have a sign at the disconnects in different locations that indicates where the other disconnects are located.

d. At the utility's discretion, you should put a sign at the main service entrance indicating where the dc disconnect is located.

63. You are connecting three different 5kW inverters to a subpanel on a residential service. Each inverter is connected to a 30A breaker on the subpanel. What is the smallest breaker that you could use for the breaker at the main service panel that is feeding the MLO (main lug only) subpanel?

a. 80A
b. 90A
c. 100A
d. 75A

64. Using the information for Question 63 above, what would be the main breaker and busbar combination of the panelboard that would work best for a load-side connection?

a. 200A main breaker, 225A busbar
b. 400A center-fed main breaker, 450A busbar
c. 400A main breaker, 400A busbar
d. 175A main breaker, 200A busbar

65. You are converting a grid-tied PV system to an ac coupled grid-tied battery backup PV system. Which of the following must you do?

a. Use 50Vdc or higher rated disconnects for 48V battery circuits.

b. Make provisions to equalize batteries if using sealed valve regulated lead-acid batteries. There shall be a schedule for equalization kept at the site of the battery bank.

c. Have a sign that indicates grounded conductor, polarity of grounded conductor, voltage, max short circuit current, and date calculations were performed.

d. You must backup all of the circuits in the building. Selective loads are not allowed due to firefighters not knowing which loads are backed up.

66. If a battery temperature sensor is not connected in New Jersey, what could be the problem with the batteries?

- **a.** Undercharge in the summer and overcharge in the winter
- **b.** Too much current in the winter
- **c.** Undercharge in the winter and overcharge in the summer
- **d.** Excessive current when it is hot

67. You are working on a project in Syria on a grid-tied 100kW project and a long piece of metal from an explosion has impaled a PV module. The array seems to be working fine and is feeding the grid, just like the day before. Of the following, which is likely the biggest problem?

- **a.** There is a ground fault.
- **b.** The metal impaling the module would likely be energized and dangerous to touch.
- **c.** There is an ac to dc short circuit.
- **d.** Lightning arrestors may have been activated.

68. The module interconnects should be:

- **a.** Anodized
- **b.** Polarized
- **c.** Galvanized
- **d.** Lubricated

69. Which of the following is the wrong way to do a supply-side connection?

- **a.** Split bolt onto service entrance conductors.
- **b.** Splicing into the conductors between the main breaker and the meter with a three-port lug.
- **c.** Bolting conductors onto busbar on opposite side of main breaker from the meter.
- **d.** Pull the meter before making the connection.

70. What are the best tools for checking performance of a system (commissioning)?

- **a.** Irradiance meter, laser thermometer
- **b.** Solmetric Suneye, inclinometer
- **c.** Tape measurer, camera, camera phone
- **d.** Solar Pathfinder, compass, level

Practice exam 2

With answers and explanations

Chapter 11

This chapter is by far the most valuable part of this book. Once you get the basics down, the most efficient way to study for an exam is by taking practice exams and going over the answers in detail. Much of the basic concepts that we use are first introduced in the *Solar Photovoltaic Basics* book. Here we are putting those basic concepts in an advanced, realistic context.

1. You are installing a concrete encased electrode (ufer) on an inverter pad. There is a conductor that is coming out of the concrete to make a connection to attach the electrode to the PV system. What is the largest that this conductor has to be if it is copper?

   a. **4AWG**
   b. 3/0
   c. 8AWG
   d. 6AWG

In the beginning of the NEC, you can look up the **contents** of the NEC and go to **Article 250: grounding and bonding** (which you should memorize) or by looking at the contents in the front of the NEC you find **Grounding electrode system and grounding electrode conductor.**

Alternatively, you can look in the index for **electrodes, grounding concrete encased** where you will be directed to different sections.

250.50 Ex. (Ex. is for exceptio)
250.52(A)(3)
250.66(B)
250.68(A) Ex. 1
250.70

250.66 gives you sizes for the ac grounding electrode conductors (GEC) and **250.66(B) gives us the sizes of the conductor that is the connection to the concrete encased electrode, to the GEC.**

The largest connection for an ac GEC for a concrete encased electrode in this case is:

4 AWG, according to 250.66(B)

For dc, the portion that is the sole connection to the grounding electrode is not required to be larger than:

4 AWG copper, according to 250.166(D)

Since we are talking about the portion that comes out of the concrete, it only has to be 4 AWG.

If we were looking at all GECs, then the answer would be 3/0, but we are just looking for the electrode where it comes out of the concrete.

From the words of the NEC, we are looking for the "connection to the GEC" of the concrete encased electrode.

2. You are building a new house in the woods with a ground-mounted PV system 1000 feet away in an opening. You decide you would like to use a ground ring, since you have a lot of 2 AWG wire left over. How deep must you bury the ground ring?

a. 18" in conduit
b. **30" or more**
c. 8 feet or more
d. As deep as the foundation

A ground ring is a type of grounding electrode. If you did not know that, you could look up ground ring in the index of the NEC.

Ground ring:

250.52(A)(4)
250.53(F)
250.66(C)
250.166(E)

We know that **250.66 and 250.166 are for sizing ac and dc GECs** (grounding electrode conductors) and since we want to know about the electrode itself, not the conductor that goes to it, let's go to **250.52 and 250.53**.

250.53 is "grounding electrode system installation" (getting close).

250.53(F) says that the ground ring must be buried at least 30".

3. The inverter went over voltage in Kalamazoo one cold winter due to global warming. The warranty was voided and the installer was responsible for buying the new inverter. The maximum input voltage of the inverter was 1000V. The PV module specifications at STC were: power = 245W, Vmp = 30.1V, Imp = 8.2A, Voc = 37.7V, Isc = 8.7A. The temperature coefficient for the monocrystalline silicon module was -0.121V/C. There were 24 modules per PV source circuit. How cold did it get when the inverter got to be 1000V dc?

- **a.** **-8C**
- **b.** 0C
- **c.** -18C
- **d.** -12C

For 24 modules in series to reach 1000V, first calculate the voltage of each module when the PV source circuit (string) reaches 1000V:

1000V/24 in series = 41.7V per module

Next, calculate the difference in voltage between STC (25C) on the datasheet and the voltage per module when the PV source circuit is 1000V:

41.7Voc at cold temp – 37.7Voc at STC = 4V change in voltage

If the change in voltage is 4V per module and for each degree C decrease in temperature every module gained 0.121V, then calculate how many degrees change in temperature is equivalent to a change of 4V:

4V change/0.121V per C = 33C° change in temperature

Since the voltage went up, the temperature went down 33C from STC (25C) so we can calculate:

25C – 33C = –8C temperature

Therefore it was **-8C** when the array was 1000V and at the maximum voltage for the inverter.

4. Four 240V 3kW utility interactive inverters are connected to a 100A subpanel on a house that is fed by a 100A breaker. The subpanel was installed for solar and there are no loads on it. What must the label on the subpanel say?

a. WARNING! INVERTER OUTPUT CONNECTION; DO NOT RELOCATE THIS OVERCURRENT DEVICE
b. WARNING! DO NOT CONNECT MULTIWIRE BRANCH CIRCUITS!
c. WARNING! ELECTRIC SHOCK HAZARD DO NOT TOUCH TERMINALS TERMINALS ON BOTH THE LINE- AND LOAD-SIDE MAY BE ENERGIZED IN THE OPEN POSITION
d. **WARNING! THIS EQUIPMENT FED BY MULTIPLE SOURCES. TOTAL RATING OF ALL OVERCURRENT DEVICES, EXCLUDING MAIN SUPPLY OVERCURRENT DEVICE, SHALL NOT EXCEED AMPACITY OF BUSBAR**

First, we will calculate the current of a 3kW inverter:

3000W/240V = 12.5A

Then we will multiply by 1.25 for continuous current:

12.5A × 1.25 = 15.6A

The continuous current of three of these inverters will be:

15.6A × three inverters = 46.8A

a. If we try to apply the 120% rule, the most continuous current we can feed a 100A subpanel with a 100A breaker is 20A (that would be 16 amps of inverter).

We could not use the 120% rule for this installation, so the 705.12(D)(2)(3)(b) label cited in a. that says:

WARNING!
INVERTER OUTPUT CONNECTION;
DO NOT RELOCATE THIS OVERCURRENT DEVICE

This label is not required.

b. If we look at b., where it says **do not connect multiwire branch circuits**, that label is for standalone 120V inverters that are using a 120/240V panelboard that is modified for 120V use only. Therefore b. is not correct. Also, part of the phrase on that label is missing. It should also mention it is a 120V supply when it is used.

c. The label referencing terminals being energized on the line- and load-side is not required to be used for the ac side of utility interactive inverters. Any UL 1741 listed utility interactive inverter will deenergize on the inverter side of an ac disconnect immediately. This is called anti-islanding and is a requirement for all grid-tied systems. c. is definitely not the correct answer (although some AHJs that are inexperienced with solar incorrectly assume that this label should be required on the ac side).

d. New in the 2014 NEC is a provision that was put in so that we could combine inverters with subpanels without being constrained by the 120% rule.

This installation would not be allowed according to the 2011 NEC and earlier, however, in the 2014 NEC, **705.12(D)(2)(3)(c)** allows us to put up this label:

WARNING:
THIS EQUIPMENT FED BY MULTIPLE SOURCES
TOTAL RATING OF ALL OVERCURRENT DEVICES,
EXCLUDING MAIN SUPPLY OVERCURRENT DEVICE,
SHALL NOT EXCEED AMPACITY OF BUSBAR

The correct answer is **d**. Fortunately, now we can add up all of the load and supply breakers (excluding the main supply) on a busbar, and as long as they do not exceed the rating of the busbar and as long as the busbar is properly protected by a main breaker, then we are good!

5. A monitoring unit is positively grounded. What would be the appropriate color coding and where do the disconnects open the circuit?

a. Grounding conductor should be white and positive and negative should be disconnected by a single switch. Overcurrent protection on negative and positive.

b. **Disconnects on negative. The negative can be black. The grounded conductor is white. Overcurrent protection on negative.**

c. Blue negative, white positive, yellow ground with green stripe, and disconnect positive and negative. Overcurrent protection on negative.

d. White negative, black positive, green ground, and disconnect opens black and white conductors. Overcurrent protection on positive and negative.

a. With a positively grounded system, we should have a white grounded conductor. We can rule out a. since a. called for a white grounding conductor. **Whenever grounding ends with a G, it should almost always be G for green or bare.**

b. is the correct answer. Just like for a positive grounded PV system, the grounded conductor is white, the other conductor (in this case negative) can be black (or another color besides green or white). The wire that is not the grounded conductor should have disconnects that open the circuit and that is also where we put overcurrent protection devices. We do not put overcurrent protection on or disconnect grounded conductors.

In summary:

Grounded conductor is white
Non-grounded conductors have disconnects and fuses.
Grounding in the name should be green or bare (EGC or GEC).

Other terms for non-grounded conductor are: hot wire, Line 1, Line 2, Line 3, ungrounded conductor.

c. can be ruled out because we do not disconnect the grounded conductor.

d. can be ruled out because we do not have a white grounded conductor and we do not put overcurrent protection or disconnects on the grounded conductor.

6. An interactive UL 1741 listed inverter is turning on and off on sunny days during what would be expected to be the time of day for peak performance. You are sent to the jobsite to troubleshoot the problem. Of the following, which would be the most likely scenario?

- **a.** Voltage too low at the ac terminals of the inverter
- **b.** **Voltage too high at the ac terminals**
- **c.** Voltage too high at the dc terminals
- **d.** Frequency too high at the terminals

Let us rule out the wrong answers first and then settle on the right answer.

a. In most places, the utility voltage does not sag significantly just during the parts of the day that the PV production would expect to peak. This is probably **not a likely scenario.**

c. During the peak part of the day for sunshine, usually around noon, we would not expect the dc voltage to get too high. If the dc voltage would get too high, we would expect that to be at the coldest time of the day and year on a system that was improperly designed. We would also not expect this problem to repeat itself on a regular basis and we could expect the inverter to break if the dc voltage was too high.

d. In most places where there are utility interactive inverters, it is uncommon for inverters to shut off due to the frequency being too high. In most of North America, for example, the frequency is so reliable on the grid that we set our clocks to it.

b. **is the correct answer**. When the conductors on the ac side of the inverter are undersized or there are other causes of resistance, we will get voltage drop. **Voltage drop on the ac side of the inverter causes the voltage to rise at the inverter terminals**, since the power is coming from the inverter, the voltage will be higher at the inverter and the voltage will drop as it approaches the interconnection. **Many people in the industry call this phenomenon voltage rise**.

7. Which is the best way to determine the state of charge of a typical flooded lead-acid battery:

- **a.** **Hydrometer testing**
- **b.** Test voltage at minimum load
- **c.** Charge battery for 25 minutes and then immediately measure voltage with a digital multimeter
- **d.** Run down battery 50% then test depth of discharge and calculate 1 – depth of discharge = state of charge

a. **is the best answer. Test the battery state of charge with a hydrometer**. A hydrometer is a device that can determine the specific gravity (density) of the battery electrolyte. A charged battery with a better state of charge has a higher specific gravity. This is because with a typical lead-acid battery, sulfuric acid is heavier than water. The specific gravity of water is 1.0 by definition.

State of charge (SOC) can also be tested with voltage readings; however, for the voltage readings to be most accurate, the battery needs to be fully charged and then to sit for a number of hours, preferably 24, before the voltage reading is taken.

b. is not the best answer, because testing voltage under load will not give you an accurate voltage reading, since the voltage will drop considerably under different loads and at maximum load, we would have trouble telling what the SOC was. Load testing can be very helpful and we can find out a lot about the health of the battery with a load test.

c. is not the best answer, because if we want to get an accurate reading, we would need to let the battery sit for at least a few hours before we took a voltage reading.

d. is a made up answer that is not practiced in the industry.

Another way of testing the state of charge of a battery is with a refractometer. A refractometer will tell how the sulfuric acid bends light different than water. The results will be much like that with a hydrometer.

Figure 11.1 Battery state of charge as related to specific gravity and battery open circuit voltage, courtesy of Trojan Battery Company 2014

Specific Gravity Corrected To	Open-Circuit Voltage			
	6V	8V	12V	24V
1.277	6.37	8.49	12.73	25.46
1.258	6.31	8.41	12.62	25.24
1.238	6.25	8.33	12.50	25.00
1.217	6.19	8.25	12.37	24.74
1.195	6.12	8.16	12.27	24.48
1.172	6.02	8.07	12.10	24.20
1.148	5.98	7.97	11.89	23.92
1.124	5.91	7.88	11.81	23.63
1.098	5.83	7.77	11.66	23.32
1.073	5.75	7.67	11.51	23.02

8. You are assigned to install a PV system with a combiner box at a car wash in an outdoor area subject to high-pressure water, air blown dust, and high temperatures. Which type of enclosure should you specify at the electrical supply house?

a. NEMA 3R

b. **NEMA 4X**

c. NEMA 11
d. UL 1703

The answer to this question likely depends if you can find **Table 110.28: enclosure selection.**

In the NEC there is no great way to look up **Table 110.28**. You can go to Article 100 definitions and look up "enclosures," where it refers you to Table **110.28**.

Also, know that it is not far after table 110.26(A)(1): working spaces.

Once you find **Table 110.28, enclosure selection** you can look through the chart and **rule out NEMA 3R, since it will not work in a location with windblown dust.**

We can see that **NEMA 4X will work under all of the specified conditions** in the question.

NEMA 11 is not in the NEC.

UL 1703 is the criterion under which PV modules must be listed and tested in the USA, not combiner boxes. (**Combiner boxes, inverters and charge controllers should be listed to UL 1741.**)

b. NEMA 4X is the correct answer.

9. You are commissioning a PV system in a humid location near salt water and you notice that in a positive grounded PV system, when the system is on the grounded conductor measures near 0V to ground and the negative wire measures 412V to ground. When you turn the system off at the dc disconnect, the grounded conductor at the combiner box stays near 0V and the negative terminal at the combiner box goes up to 494V. What is the problem with this situation?

a. Negative ground fault at array
b. Positive ground fault at combiner box
c. System ground has come loose and is shorting intermittently
d. **System may be within normal limits**

In a positively grounded system, the measurement of voltage with positive to ground should be very close to 0V. Sometimes the voltage will drift a little way from 0 due to voltage drop.

Also in a positively grounded system the negative should operate near maximum power voltage (Vmp) when the system is producing power and the negative (ungrounded) conductor should operate near Voc when the system is turned off.

The example here indicates that nothing is wrong with this PV system.

b. is the correct answer, since nothing is out of normal limits.

10. There is a PV system that is not working and you are sent out to troubleshoot the problem. The inverter is about 4 years old and is the type with a ground-fault detection and interruption fuse (GFDI). The inverter is a grounded inverter. The light is blinking that indicates a ground fault is present. You are about to go on the roof and you just realized that you do not have fall protection with you. Also, you notice that the dc equipment grounding conductor was removed from the inverter by one of your coworkers that tried to troubleshoot the system the day before. What might be a hazard with this system?

- **a. The array frames and racks may be energized with reference to ground.**
- **b.** Since the grounded conductor cannot be disconnected, at least you know that the white wire is at the same potential as the equipment.
- **c.** It is likely that the array voltage is 0 on positive and negative grounded because of a short circuit.
- **d.** The green and the white wire will have the same voltage.

This system could be dangerous and the installer should definitely not go on the roof without fall protection. Additionally, you should not work on this job alone.

When a ground fault is indicated, normally grounded conductors can be energized with reference to ground. This means that the white wire can be energized with reference to the green wire. **That means we can rule out answer d.**

The reason that the white and green wires are at a different potential after a ground fault is that the **(GFDI) fuse that is linking the green and white wires has opened and the green and white wires are no longer linked.** This means that the **green and white wires can be disconnected** and from there **we can**

rule out answer b. (There is talk that in a future version of the NEC that we will not have white wires on the dc side of an inverter for this reason, since they are not "solidly" grounded.)

It is possible that there is a short circuit between negative and positive with 0 voltage, but this scenario is unlikely to occur. When we are taking a multiple choice exam, and there are two different possibilities, always pick the more likely answer.

a. is the correct answer. The array frames and racks may be energized with reference to ground. We mentioned that the dc equipment grounding conductor was removed from the inverter. This makes it very likely that the array frames and racks could be energized and it makes a dangerous situation where someone could be shocked. The equipment grounding conductor should always be connected to ground!

11. You are required to put posts into concrete and the specifications call for holes that are 18" diameter and concrete going 5 feet deep. The holes are going to have 3" diameter pipe going 5 feet down. If you are going to have 188 of these posts, then how many cubic yards of concrete do you need?

a. 1617 yards
b. 64 yards
c. **60 yards**
d. 102 yards

There are many ways to do these calculations. We can start off working in inches or feet. If we start in inches, then we will determine the area of the hole as an area of a cylinder. The area of a circle is calculated by the formula:

$$3.14 \times \text{radius}^2 = \text{area}$$

The width of the hole was the diameter and the radius of a circle is half of the diameter. And since the diameter is 18", then the radius is 9".

$$3.14 \times 9^2 = 254 \text{ in}^2$$

Then to get the volume of the hole we will multiply the area of the circle by the depth of the hole in inches.

Depth of hole is 5 feet × 12" per foot = 60"

$$60'' \times 254 \text{ in}^2 = 15{,}240 \text{ in}^3 \text{ per hole}$$

Since we are putting posts into the holes, we can subtract the volume of the posts from the volume of the concrete. We will figure out the volume of the pipe with the same formulas.

Since the diameter of the pipe is 3", then the radius is 1.5":

$$3.14 \times \text{radius}^2 = \text{area}$$
$$3.14 \times 1.5^2 = 7 \text{ in}^2$$

Then we multiply by the depth of the pipe where the pipe will be displacing the concrete:

$$5 \text{ feet} \times 12'' = 60''$$
$$60'' \times 7 \text{ in}^2 = 420 \text{ in}^3 \text{ of displaced concrete per hole}$$

Then we will subtract the volume of displaced concrete from the volume of the hole:

$$15{,}240 \text{ in}^3 - 420 \text{ in}^3 = 14{,}820 \text{ in}^3$$
needed per hole

Next, we will determine the conversion factor from cubic inches to cubic feet by using the formula:

$$\text{Length} \times \text{width} \times \text{depth} = \text{volume}$$
$$12'' \times 12'' \times 12'' = 1728 \text{ in}^3 \text{ per cubic foot}$$
(this is a conversion factor)

Then we convert to cubic feet:

$$14{,}820 \text{ in}^3/1728 \text{ in}^3 \text{ per foot} = 8.6 \text{ cubic feet}$$
needed per hole

Since we have 188 holes to order concrete for, then we must remember to multiply by 188 holes (many people forget to do the easy part on a test, so do not forget).

$$8.6 \text{ cubic feet} \times 188 \text{ holes} = 1617 \text{ cubic feet of concrete}$$
required for 188 holes

Then we can convert cubic feet to yards and since there are 3 feet in one yard then the cubic feet in a yard are calculated by the formula:

$$\text{Length} \times \text{width} \times \text{depth} = \text{volume}$$
$$3 \text{ feet} \times 3 \text{ feet} \times 3 \text{ feet} = 27 \text{ cubic feet per cubic yard}$$
(in construction a cubic yard is called a yard)

We know that since yards are smaller than feet, the conversion from feet to yards is going to make our number smaller, so we will divide:

1617 cubic feet/27 cubic feet per yard = 60 yards
The answer is 60 yards.

Note: Author has memorized pi to 100 digits using the book *Quantum Memory*, by Dominic O'Brien. Memory techniques are a great way to assimilate information and pass tests.

12. For a large ground-mount system every eight modules will need two ballast blocks. Each ballast block is made from concrete that weighs 140lbs per cubic foot. The dimensions of each block are 2 feet 3" north to south, 3 feet 2" east to west. If the ballasts for the section of eight modules have to weigh 3824lbs, to keep the array stable, then how deep must we pour the concrete into the forms?

a. 23"
b. 46"
c. 72"
d. 48"

First, we will calculate how many cubic feet of concrete will weigh 3824 lbs:

3824lbs/140lbs per cubic foot = 27.3 cubic feet

Now knowing that **length × width × depth = volume**, and we have length, width and volume, we just need to get the units correct and then solve for volume.

We can do the equations in inches or feet and since last time we worked with inches, this time we will work with feet (inches may be easier since there are 12" per foot, which does not do well with moving decimals).

Length = 2 feet 3" = 27"
27"/12" per foot = 2.25 feet
Length = 2.25 feet

Width = 3 feet 2" = 38"
38"/12" per foot = 3.17 feet

Depth is the unknown:

Volume = 27.3 cubic feet for two ballasts

Volume for ONE BALLAST is:

$$27.3 \text{ cubic feet}/2 = 13.7 \text{ cubic feet}$$

$$(L \times W) \times D = V$$
$$D = V/(L \times W)$$

$$\text{Depth} = 13.7 \text{ cubic feet}/(2.25 \text{ feet} \times 3.17 \text{ feet})$$
$$\text{Depth} = 1.92 \text{ feet}$$

$$1.92 \text{ feet} \times 12'' \text{per foot} = 23'' \text{ deep}$$

13. On a ballasted ground-mounted PV system, the width of the footing would be wider due to:

- **a.** **Soil load bearing**
- **b.** Frost line
- **c.** Wind uplift
- **d.** Insulation

There may be a few answers here that can be argued, but the best answer appears to be soil load bearing. If the ballast was narrow and the soil was soft, it would sink into the soil.

14. Fidelity Roof Company is a third generation roofing company with a reputation for quality. The roof being looked at for solar has been well taken care of by three different generations of roofers and each generation has put another roof on the building. What would be the main concern to look for when determining if solar should go onto this composition asphalt shingle roof?

- **a.** The buildup of roofing materials is too thick for a good attachment.
- **b.** Redwood lumber will get too weak over the span of 60 years.
- **c.** **Too much weight from layers of roofing.**
- **d.** There are no good waterproofing materials compatible with asphalt shingles.

Having the roof too thick can be a problem, but we can make hardware that can penetrate the thickest residential roof.

Oftentimes, roofers will put multiple layers of roofs on top of each other. Usually after two or three layers, the roofers will have to do a tearoff and take the roofing layers off, so that the roof can support the weight of the new roof. A tearoff is a rough job. As a solar installer, you should be aware that **roofs with two or three layers on them might not be able to take the extra weight of a PV system**. It would be a decision that the permitting agency (AHJ) or an engineer might enforce. Many PV systems weigh about as much as a layer of asphalt shingle roofing materials, which is about 2.5 to 3.5lbs per square foot.

There are plenty of materials that are made for waterproof flashed attachments to asphalt shingle roofs.

Redwood lumber will last a long time, perhaps until it burns down or is destroyed another way. One problem with old houses, however, is that the lumber is often not large enough and would require reinforcements before solar could go on the house. If you see 2 × 4" lumber rafters, you can expect that you will have to reinforce the roof before installing solar.

15. The owner of a house complains of low performance. You come to the house and try to figure out what the problem is. What would be the best thing to do first of the following?

- **a.** Test the current of each string
- **b.** Test the voltage at the inverter
- **c.** Make sure connectors under modules are secured
- **d.** **Look at the grounding**

All of the above answers are good ideas when checking a system, but always think of doing the safest, most non-invasive thing first. Usually, the safest thing to do first would be talking to the customer or looking at the system before touching it.

If there were a problem with the grounding, even if it was not the cause of the low performance, it is best to address the grounding system first, so that you do not get hurt after you start touching things.

There is a safe order when checking a system:

1. Look and ask (before you touch).
2. Inspect your equipment (make sure the meter works).
3. Turn equipment off (disconnects open).
4. Test using all types of safety equipment (fall protection).
5. Make corrections (with gloves).

When you are taking a test, always err on the side of safety.

16. A giant 100 foot tall American flagpole with a giant flag waving side to side is directly south of a solar array at 38° latitude. At solar noon on winter solstice, the solar elevation angle is 28°. You are going to put solar on a rooftop and would like to make sure that the flagpole does not shade the array. Your rooftop is 65 feet high and the roof is a relatively flat roof. How far away from the flagpole should the solar array be located?

- **a.** 45 feet
- **b.** **66 feet**
- **c.** 73 feet
- **d.** 10 yards

First, we will visualize what we are dealing with. I recommend drawing a sketch.

Figure 11.2 Flagpole at PV factory at solar noon, courtesy of Sean White

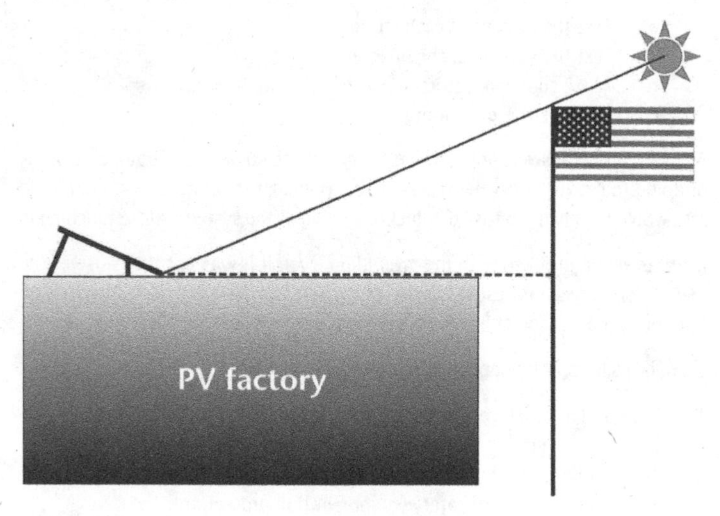

Then we will visualize the triangle and the sides and angle that we know something about or that we want to know something about.

Figure 11.3 Flagpole at PV factory at solar noon with sides and angle labelled

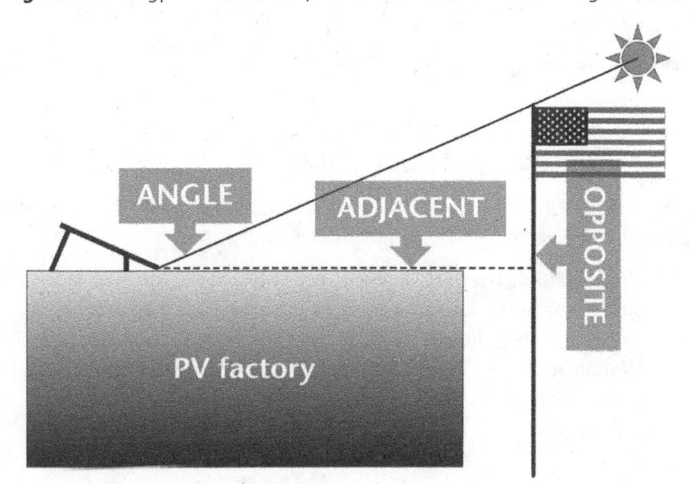

In this case, we know the angle, which is the angle of the sun. We can easily figure out the opposite side of the triangle by subtracting the building height from the flagpole height.

Flagpole - building = opposite
100 feet - 65 feet = 35 feet
Opposite = 35 feet

If we remember the trig functions SOH CAH TOA, we can see that we have a case where we know the opposite and the adjacent sides of the triangle, so that leads us to use TOA.

TOA stands for:
Tangent = opposite/adjacent

Since we are going to solve for the adjacent side of the triangle we can do some algebra:

$$T = O/A$$
$$O = TA$$
$$A = O/T$$

To get the tangent of 28° on your scientific calculator, press **28 Tan** and you get 0.532.

Plug in the numbers:

$$A = O/T$$
$$A = 35 \text{ feet} / 0.532$$
$$A = 65.8 \text{ feet}$$

The answer is **b**. We want to put the solar array at least 65.8 feet from the flagpole to avoid noontime shading on the winter solstice.

17. You ran across a free 10kW inverter and you only have 3kW of PV to install. Since the inverter was free, you decided to install it on your house. Your rooftop can barely fit 3kW on it and there is no way you will ever put more PV on that inverter. What is the size of the breaker you will use for that system? Assume that the inverter is 240Vac.

a. 15A
b. 20A
c. 30A
d. **60A**

If you thought that you should be able to size the inverter based on the PV, you are wrong. The inverter breaker is always sized based on the inverter output current.

For this example, we will first determine the inverter output current:

$$\mathbf{10kW = 10{,}000W}$$
$$\mathbf{VI = W}$$
$$\mathbf{W/V = I}$$
$$\mathbf{10{,}000W / 240V = 41.7A}$$

Then to size the inverter breaker, we multiply by 1.25 and round-up to the nearest common sized breaker.

$$\mathbf{41.7A \times 1.25 = 52.1A}$$
Round-up to a 60A breaker.

18. You are asked to connect on the supply side of a 2000A breaker. What should you do in order to connect this system?

a. Pull the meter while wearing an arc-flash suit
b. Disconnect the current transducers

- **c. Open utility disconnects**
- **d.** Wear PFAS and steel-toed boots

Working with 2000A is dangerous. You would never want to pull the meter to disconnect a 2000A service. On a residential service, the current will go through the meter and on a service of this size, the meter will typically not have 2000A running through the meter and the meter will work with CTs (current transducers) that work like a clamp on amp meter. Disconnecting the CTs will not make anything safer to work on.

Wearing PFAS (a personal fall arrest system) cannot hurt, but is not necessary when working close to the ground.

Wearing steel-toed boots could possibly contribute to a shock hazard.

The best answer is to open the utility disconnects and you would want to work with the utility on this.

19. If you had an off-grid system with a 10kWh battery bank, a ground rod, an ac coupled inverter system, an inverter that is connected to the inverter with a 300Adc disconnect, and a 250A OCPD between the inverter and the battery bank, what would be the correct size for a copper equipment grounding conductor between the battery bank and the inverter?

- **a. 4AWG**
- **b.** 6AWG
- **c.** 250kcmil
- **d.** 1/0AWG

Look to Table 250.122 and you will see that the equipment grounding conductor (EGC) is sized from the OCPD, which in this case was 250A and leads us to use a 4AWG EGC.

Also, in 250.122, it says that the EGC shall not be required to be larger than the conductors supplying the equipment, which is not relevant here, but is good to know.

20. You are working on a jobsite and the wrong inverter was sent, because the old inverter was thought to be less safe and less efficient by the business owner. What must you do to make sure that the new ungrounded inverter is installed correctly when adapting the plan from the grounded inverter?

a. Switch to a larger grounding electrode conductor.
b. **Not use 90°C rated USE-2 for your source circuits on the roof outside of conduit.**
c. Tap the transformer on the ungrounded inverter.
d. Switch the white wire from the negative to the positive.

When switching from a grounded inverter to an ungrounded inverter there are a few differences.

A grounded inverter has a white grounded conductor on the dc side. An ungrounded inverter has no white wire on the dc side. Also, only ungrounded conductors go through fuses or poles of disconnects. If there is an ungrounded inverter, there are twice as many ungrounded conductors, and twice as many fuses and poles of disconnects that conductors have to go through.

The correct answer here is **b.**, to not use USE-2 wire. PV wire must be used on ungrounded inverters for the module interconnects and wherever PV source circuits are not in conduit.

21. A 1MW inverter with 1200kW of PV is operating on a sunny day with 780W per square meter of irradiance and a cell temperature of 45°C and dc to ac derating of 95%. What is the expected inverter output power? Assume a typical temperature coefficient of power of -0.45%/C

a. 1MW
b. 1.1MW
c. 908kW
d. **809kW**

We are looking for different derating factors and we can see three of them here:

1. Irradiance below 1000W/square meter
2. Cell temperature above 25°C
3. Given derating of 95%

We need to get derating numbers for these factors:

For irradiance, if we have 780W/square meter out of 1000W/square meter

780/1000 = .78 derating for irradiance

For temperature, we are going to lose some power.

First, we are going to determine our difference in temperature from how the modules were tested, which was 25°C.

45°C – 25°C = 20°C

Then we multiply our difference in temperature (delta T) by the temperature coefficient for power.

20°C × –0.45%/°C = –9% change in power

Since we lose 9% out of 100% then:

100% – 9% = 91%
Change to decimal 91%/100% = .91
91 derating for temperature

Our last derating was a given 95% or .95.

Now we multiply our PV capacity of 1200kW by the derating factors:

1200kW × .78 × .91 × .95 = 809kW

We should double check that the 1MW inverter can handle the output of 809kW, which it can without clipping.

Therefore, we would expect this system to get somewhere near **809kW output** during these conditions when there are no problems.

22. You are studying notes from a site survey and note that the panelboard busbar is rated at 425A and the main breaker is rated for 400A. The service voltage is 240Vac. There are 500A of load breakers connected to the panelboard. What would be the size of the largest ungrounded inverter that you could connect on the load-side?

- **a.** 110kW
- **b.** **21kW**
- **c.** 33kW
- **d.** 26.4kW

When doing a load side connection, if you can put the main breaker on the opposite side of the busbar from the main breaker, then you can put 20% more on the busbar than you already would have been able to do if the source (or sources) were all feeding the same portion of the busbar. This is commonly referred to as the 120% rule or 705.12(B)(2)(3)(b). (The sum rule, aka 705.12(D)(2)(3)(c) could not apply here due to the existing 500A of loads.

Figure 11.4 120% rule with main and solar breakers at opposite sides of busbar

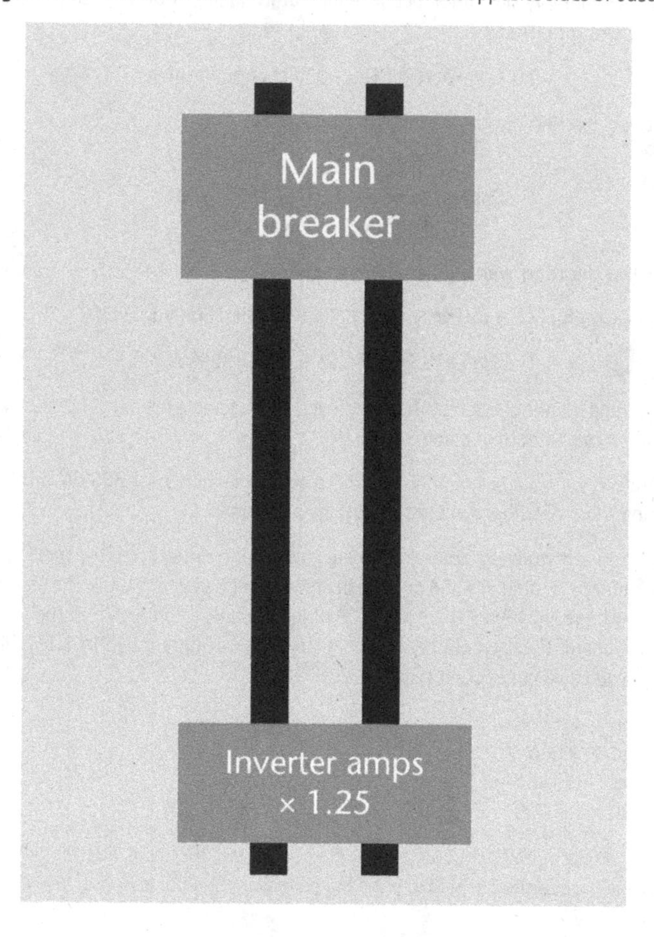

The 120% rule states that 125% of the inverter current plus the current of the main breaker can be no more than 120% of the rating of the busbar as long as the breakers feeding the busbar are at opposite ends of the busbar.

$$\text{(Inverter current} \times 1.25) + \text{main breaker current} \leq \text{busbar} \times 1.2$$

We can also do some algebra and make the equation look like:

$$\text{Inverter current} \leq ((\text{busbar} \times 1.2) - \text{main breaker})/1.25$$
$$\text{Inverter current} \leq ((425\text{A} \times 1.2) - 400\text{A})/1.25$$
$$\text{Inverter current} \leq 88\text{A}$$

Therefore, the most current that we can come out of the inverter in this case is 88A. Since the ac voltage in this case is 240Vac, then:

$$240\text{V} \times 88\text{A} = 21{,}120\text{W}$$
$$21{,}120\text{W}/1000 = 21\text{kW}$$

If you studied the 2011 NEC or earlier, the 120% rule used to use the inverter breaker in calculations. Big changes in Article 705 in 2014 caused us to use the inverter current × 1.25, which could give us a little extra inverter in some situations due to the fact that inverter breakers are sized by multiplying inverter current by 1.25 and then rounding-up to the next common breaker size (if under 800A). It is this rounding that we do not have to consider, that may give us some more inverter.

23. When using a safety monitoring system, a designated safety person must

- **a.** Be in charge of installing a line with flags a minimum of 6 feet apart on a line.
- **b.** Inspect fall protection systems and run weekly toolbox meetings when there are groups of 11 employees or more.
- **c.** **Have no other duties, be on the same level of work and close enough for oral communication.**
- **d.** Be present for all work over 6 feet in height.

A safety monitoring system is used when a designated safety person watches out for employees and tells them when they get too close to the edge or in a dangerous situation. The designated safety person must be competent to identify safety hazards. They must also be on the same level as the people they are watching, be able to verbally communicate with those who they are watching out for.

On low slope (flat) roofs with a width of 50 feet or less, a **safety monitoring system** alone may be used. On roofs wider than 50 feet, a **warning line system** must be used with a safety monitoring system.

A safety monitoring system can be used without other methods of fall protection (this is often used as an excuse when roofers are caught not using fall protection).

This information can be found on the OSHA website by doing a search for 1926.502(h) and **fall protection can be found at 1926.502**. Although there is an overwhelming amount of information on the OSHA website (insomnia cure), it is a good idea to read the information on fall protection.

A great way to learn about safety is by taking a class in person or online and earning your OSHA 10 or OSHA 30.

24. You are called to a jobsite and troubleshooting an inverter that is not sending power back to the grid. You were told by a coworker that inside of a dc disconnect that the white wire to ground is 402V and the black wire to ground is less than 1V. What is the most likely problem?

- **a.** The inverter is turned off and just needs to be turned on by closing the dc disconnect.
- **b.** There is a ground fault on the grounded conductor.
- **c.** **There is a ground fault on the ungrounded conductor.**
- **d.** There is a positive to negative short circuit.

Figure 11.5 Ground fault label for grounded inverter, courtesy of pvlabels.com

According to Section 690.5(C), the label shown in Figure 11.5 should be placed on a utility interactive inverter of a grounded PV system. (An ungrounded PV system will not have an ungrounded white dc wire.)

When there is a ground fault on the ungrounded conductor, the ground-fault detection and interruption (GFDI) fuse will open up, thereby disconnecting the grounded conductor from ground and making the white, grounded conductor ungrounded.

In this same scenario, since there is a ground fault on the ungrounded conductor, it would make the voltage of the ungrounded conductor the same as ground.

Let's say there is a negatively grounded PV system with a ground fault on the positive conductors going back to the inverter. This would cause the given scenario.

Figure 11.6 shows 5 different locations for a ground fault. The ground fault in the question that we are analyzing is in position number one.

Figure 11.6 PV source circuit possible ground fault locations

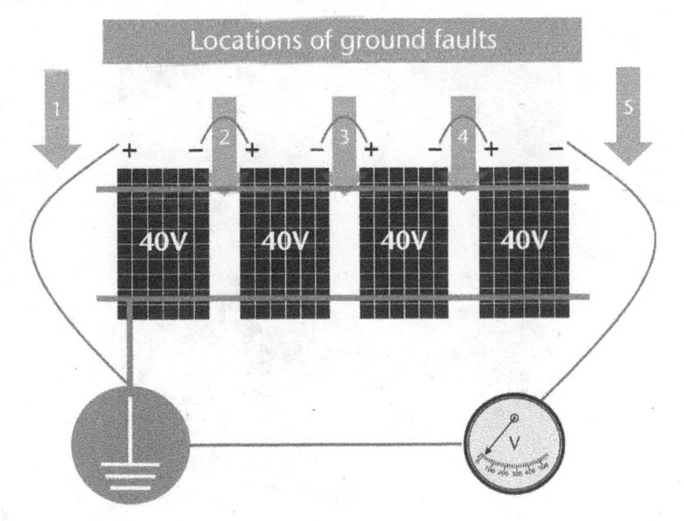

Table 11.1 Detecting location of ground fault with voltage test

Negative to ground	Positive to ground
160V	0V
120V	40V
80V	80V
40V	120V
0V	160V

We can see that if the ground fault were in a different position, that the voltage to ground would be different. Using this information, you should be able to troubleshoot ground faults from a combiner and determine where on a string the ground fault is located. This is good detective work.

Be sure that if you are troubleshooting ground faults from a combiner that you first open the disconnect at the combiner, determine that there is no current on the strings and then carefully, using gloves and proper PPE (personal protective equipment), open up the fuses, which are typically in touch safe fuse holders.

Figure 11.7 Touch Safe Fuse Holder, courtesy of Schurter, Inc.

Touch safe fuse holders work as a non-load-break-rated disconnect. Never open fuses when they are under load (there is current); they could catch fire! Dc arc-fault protection can reduce this risk.

When troubleshooting a combiner, if you do not open the fuse connections between the PV source circuits and the combiner busbar, then the voltages will all be the same as the busbar and you cannot isolate source circuits (strings) to check for ground faults.

25. A broken fuse grounded inverter has been replaced with a newer non-isolated (aka ungrounded) inverter. You were not the one to replace the inverter. Six months after the inverter has been replaced, the monitoring system has determined that the system suddenly has stopped producing energy. Your supervisor has checked everything and said he cannot find the problem. He said that voltage to ground on the white dc wire was0V. What else would you expect to find and what would you expect the problem to be?

- **a.** **There is a ground fault and the equipment would be expected to be the same voltage as the white wire in this scenario.**
- **b.** There is a short and the ungrounded conductor will be at the same voltage as ground until the problem is fixed.
- **c.** There is a short and the ungrounded conductor will be at the same voltage as the inverter cabinet until the problem is fixed.
- **d.** There is a ground fault on the ungrounded conductor and the white wire is therefore ungrounded by the fault.

Fuse grounded inverters were supposed to be installed with a white wire in the 2014 NECC and earlier. In the 2017 NEC, we will install fuse grounded and non-isolated inverters the same way and will call them both functional grounded systems. Most fuse grounded inverters will be replaced within the next 10 years at the end of their lifespans. When replacing these inverters, we need to either replace the white wire or mark the white wire, which should be acceptable to most AHJs. According to the 2014 and earlier versions of the NEC, we would have needed to replace USE-2 wire with PV wire on PV source circuits, but this is no longer the case after the implementation of the 2017 NEC where USE-2 wire is acceptable for all inverters.

Non-isolated inverters will typically operate where there is an electronic means within the inverter to keep the ground voltage right in between positive and negative. If ground voltage is 0 to positive or negative in this scenario, then it would indicate a ground fault (unless it was "round midnight").

If the new non-isolated inverter was working for 6 months, then it was more than likely that the white wire was ungrounded. Since the white wire was grounded when it was tested, then there was a ground fault on the white wire (that should have not been white).

Assuming that all of the equipment was properly bonded, then it should all be at the same potential and grounded with reference to earth.

Test psychology: If we read carefully answers b. and c., we can see that they are really the same answer. Since a multiple-choice test of this type cannot have two correct answers, then we can rule out both answers. Then it just becomes a choice between a. and d. Answer d. can be ruled out, since the white wire is grounded by the ground fault.

26. What is the safest way to connect load- and line-sides of ac and dc disconnects on a utility interactive system?

- **a.** Load-side of dc disconnect to inverter and line-side of ac disconnect to inverter
- **b.** Line-side of dc disconnect to inverter and load-side of ac disconnect to inverter
- c. **Load-side of dc disconnect to inverter and load-side of ac disconnect to inverter**
- **d.** Line-side of dc disconnect to inverter and line-side of ac disconnect to inverter

Disconnects, meter bases, and other electrical equipment are typically more protected on the line-side. As we can see in the image above, the upper portion of the disconnect is protected by plastic, where the lower portion has screw terminals that can easily be touched. The protected side is where we are better off putting the more dangerous conductors.

On a dc disconnect we know from the sign that line- and load-sides of the disconnect can be energized in the open position. However, the more dangerous side is the side of the disconnect that goes to the PV. **The line-side of a dc disconnect in an interactive PV system should be connected to the PV.**

Figure 11.8 SMA combi-switch, courtesy of SMA

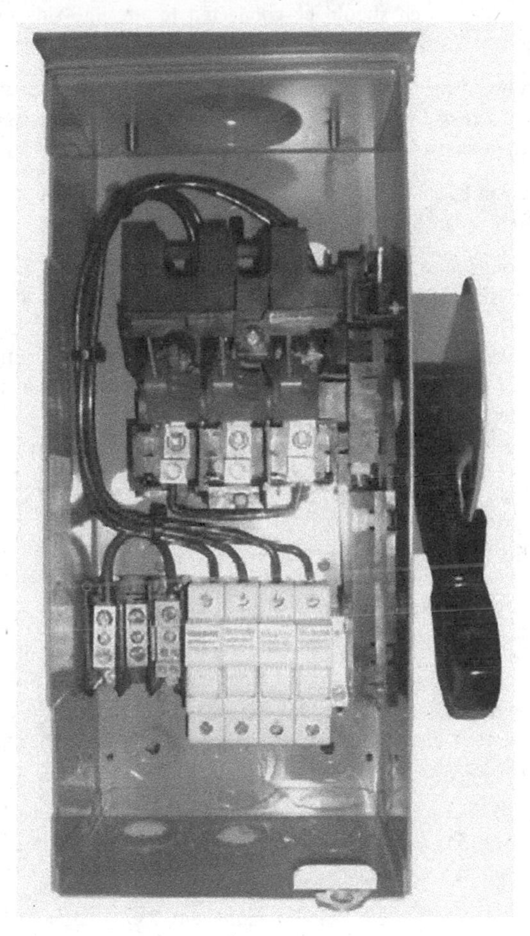

The ac disconnect is more interesting. Due to anti-islanding provisions of the grid-tied inverters, when there is an open circuit, the inverter side of the ac disconnect will immediately be deenergized and the utility side of the disconnect will have grid power and potential high fault currents coming to the ac disconnect. Therefore, **the more dangerous side of the ac disconnect is the utility side, which should be connected to the line-side of the ac disconnect.**

A simple way of remembering this is to remember that **the inverter should be on the load-side of the ac and the dc disconnects.** Think of the electronics in the inverter, making it safer with technology.

27. There is a building that you would like to put different inverters on for different tenants. What would be required in this situation?

- **a.** There must be six or fewer PV disconnects to turn off all of the PV systems and all of the systems must be in the same enclosure.
- **b.** There must be 12 or fewer PV disconnects.
- c. **There shall be a maximum of six disconnects to turn off all of the PV systems and the disconnects need to be in the same location.**
- **d.** There may be an unlimited number of disconnects as long as they are the anti-islanding type.

Part III of Article 690 of the NEC is titled "Disconnecting means." If we look a little farther down the page, we see **690.13(D): maximum number of disconnects,** where it says that we cannot have more than six disconnects in an enclosure or **group of enclosures.**

This means that we need to be able to turn off the PV systems with no more than six different switches to turn off that are PV switches. Since it says that the switches can be in a group of enclosures, then it would be acceptable to have the switches in different enclosures as long as they are in the same area.

690.13(D) also says that a single disconnecting means can disconnect multiple interactive inverters. So, we can definitely have more than six inverters on a building, if you were wondering.

As a side note, in **Article 230: service equipment,** 230.71 (often called the six handle rule) states that there should be no more than six service disconnects per service. It is generally understood in the PV industry that PV is not a service and that you can have six service disconnects in addition to six PV disconnects, however, there will be some inspectors that will disagree with this interpretation.It is a best practice to try to locate all of the PV disconnects and the service disconnects in the same location. If everything is not in the same location, you should have a sign at each location telling where everything else is located.

This is so that someone unfamiliar with the site will not miss something when they are trying to turn everything in the building off. This is especially helpful for the fire department.

28. A PV dc disconnect may go in any room in the house except for the:

- **a.** Kitchen
- **b.** **Bathroom**
- **c.** Bedroom
- **d.** Nursery

690.4(E) says that the PV system **disconnecting means and PV system equipment** are not allowed in the bathroom.

How do we remember that we cannot have the disconnect in the bathroom? Think of wet feet not being compatible with electricity. Sometimes it may be convenient to turn off your PV system when you are doing business in the bathroom, but you are not able to under any circumstances.

29. What are the rapid shutdown voltage limits for inside the array boundary and outside the array boundary?

- **a.** 30V in 30 sec inside and outside the array
- **b.** 30V inside array and 80V outside of the array
- **c.** **80V inside the array and 30V outside of the array**
- **d.** 80V in 30 sec inside and outside of the array

690.12(B)(1): outside the array boundary calls for 30V within 30 seconds.

690.12(B)(2)(2): controlled conductors inside of the array boundary calls for 80V within 30 seconds

Regarding inside of the array boundary, there are other uncommon possibilities, such as 690.12(B)(2)(3), which was instituted for BIPV, where you get away from the 80V rule and go up to 600V on one and two family dwellings and 1000V on other buildings if there are no exposed connective metal parts or exposed wiring.

30. You are using 270W microinverters and 300W solar modules on a residential service at 240Vac. You want to make sure that you are using the right number of microinverters for the cable that the microinverters are connected to by using the NEC. If the cable is capable of handling 25A, then how many microinverters can you have on the cable?

a. 25
b. **17**
c. 18
d. 22

For this calculation, we will start by using the microinverter output power. The power of the PV module is not used in this calculation. Whenever you are doing calculations for the output of an inverter, do not use the power of the PV modules in the calculation.

First, we want to determine the output current of each inverter:

Volts x amps = watts
Watts/volts = amps
270W/240V = 1.125A

To determine the overcurrent device size for an inverter, we multiply output current by 1.25:

1.125A × 1.25 = 1.41A

Then we see how many microinverters can be on a 25A cable:

25A/1.41A = 17.7 microinverters

We have to round-down to stay safe, i.e., **17 microinverters.**

31. You are installing a PV system on a building and the plans call for AC cable in a wet location. What is the best solution?

a. Use AC cable with raintight connectors.
b. **Change to EMT.**
c. Use AC cable with weathertight connectors.
d. Use PV wire inside of AC cable.

In the NEC contents, look under **Chapter 3: wiring methods** and find **Article 320: armored cable, type AC**. This is where you can look up the different types of cable, conduit, and wiring methods.

Alternatively, you can use the index of the NEC and look up AC cable to get you to the same place.

When you start reading Article 320, you will soon discover that AC cable may not be used in wet locations, therefore if you install it in a wet location, that would not be code compliant, regardless of what kind of connectors were used.

To double-check that EMT (electrical metallic tubing) is the correct conduit and wiring method, we can use the contents or index to look for EMT in Chapter 3: wiring methods. We will find **Article 358: electrical metallic tubing, type EMT**. We see here that EMT can be used in wet locations, which is one reason why EMT is so widely used in the PV industry. EMT is a metal raceway and can be used for dc PV circuits inside a building.

AC, dc, and MC

There has been confusion with these terms. First of all, AC can mean alternating current. In the NEC, ac is not capitalized when used for alternating current, unless it is at the beginning of a sentence or in a title. The same goes for dc and direct current.

AC cable is a cable assembly surrounded by metal and cannot be used outside.

MC cable is a cable assembly that looks like AC cable from the outside, however, we can use MC cable for PV source and output circuits inside a building. Also, MC cable can be used in wet locations.

DC PV source and output circuits within a building up to the first readily accessible dc disconnect have to be in a metal raceway or can be in MC cable.

The most popular type of connector used on PV modules is called an MC connector. This is a brand name and here MC stands for Multi-Contact.

32. What is the smallest size of EMT that you can use for 12 current-carrying 12AWG THWN-2 conductors and one 12 AWG equipment grounding conductor?

a. 1"
b. **¾"**
c. 1.5"
d. 1.25"

To determine the number of conductors that will fit into conduit, we will use the Chapter 9 tables of the NEC.

In the contents, looking at **Chapter 9**, we will use:

- **Table 5:** dimensions of insulated conductors and fixture wires
- **Table 4:** dimensions and percent area of conduit and tubing

First, we will find out the **cross-section area of a 12 AWG THWN-2 conductor in Chapter 9 Table 5, which is 0.0133 in^2**:

Table 11.2 From NEC Chapter 9 Table 5

Type	Size	Area (square inches)
THHN	10AWG	0.0211
THWN-2	12AWG	0.0133

Since we have 13 12 AWG THWN-2 conductors (12 plus a ground) then we can determine the total cross-section area of conductors to be inside of the conduit:

13 conductors × 0.0133 in^2 per conductor = 0.1729 in^2 total

In Chapter 9, Table 4, we look for EMT (Article 358 EMT).

Since we have more than two wires (we have 13), then we can use the column for 40% and use 40% of the total area of the EMT conduit.

0.1729 in^2 is between 0.144 for ½" conduit and 0.203 for ¾" conduit, so ¾" EMT would be the smallest conduit that we would use.

Table 11.3 Dimensions and percent area of conduit for EMT from NEC Chapter 9 Table 4

Trade size	Over 2 wires 40%	Total area 100%
½ inch	0.122 in^2	0.304 in^2
¾ inch	0.213 in^2	0.533 in^2
1 inch	0.346 in^2	0.846 in^2

The correct answer is to use ¾" EMT.

33. There is a new LEED latinum building going up and there is a 1000A three-phase service that you are asked to connect solar to. There are multiple 200A subpanels in the building. What would be the most likely way you would connect a 5kW system to this building?

- **a.** Load-side connection on a subpanel
- **b.** **Supply-side connection**
- **c.** Load-side connection on main panel
- **d.** Feeder tap

On new buildings with 1000A services, ground-fault breakers will be used. These breakers usually are not able to be backfed.

705.32: ground-fault protection. Where ground-fault protection is used, the output of an interactive system shall be connected to the supply-side of the ground fault protection.

There are exceptions to 705.32, but, in general, most 1000A breakers cannot be backfed. If the breaker is marked line and load, it definitely cannot be backfed.

The best answer would be a connection that is to the supply-side of the main service disconnect.

34. With a 120Vac inverter on a house with a standalone PV system, can you use a 120/240V service panel and if so, what must you do?

- **a.** Yes, as long as you do not connect Line 1 to Line 2.
- **b.** **Yes, as long as you do not use multiwire branch circuits.**
- **c.** Yes, if you design the system to not overload the neutral.
- **d.** No.

Look in 710.15(C): single 120-volt supply. 710 is standalone systems (formerly Section 690.10: standalone systems). Here it tells us that we can use 120/240V service equipment as long as we **do not add multiwire branch circuits**. There also has to be a sign that says to not add multiwire branch circuits.

Also, the rating of the inverter output overcurrent device must be less than the rating of the neutral bus. Inverter output overcurrent device must be located at the inverter.

35. You are supervising a utility-interactive job using grounded inverters and there are two different electricians installing dc disconnects. The first electrician is installing the disconnects with the green ground screw that comes with the disconnect, connecting the grounded conductor bus to ground and the other electrician was not using the green ground screw. Which electrician was right and why?

- **a.** The first electrician because, according to 110.3(B), equipment should be used according to directions.
- **b.** The first electrician because the grounded conductor needs to be grounded.
- **c.** **The second electrician because using the screw would cause a ground fault.**
- **d.** The second electrician because it is a grounded inverter.

Grounded utility-interactive inverters ground the system only in one place and that place is inside the inverter. If grounding is done in two places, then there would be currents that would otherwise run in the white colored grounded conductor, that would now also run in the green or bare equipment grounding conductor and currents could also be running through the rails and metal equipment.

System grounding should only be done in one place per system, which in a grounded system that is inside the inverter, typically through a 1A fuse (or larger with large systems). There should never be multiple places where the system is grounded. Also, on opposite sides of a transformer, there are separately derived systems.

Separately derived systems will have separate system grounding. The ac side of the system is grounded at the main service panel where the white and green wires are connected, often at a common busbar.

36. What is the best type of wire to use for series connections in a battery bank?

a. DLO
b. Heavy duty arc welding cable
c. **THWN**
d. Bare copper

When batteries are connected to a PV system in a building, the NEC will have to be followed. We do not use diesel locomotive cable (DLO), welding cable, or automotive cable for electrical systems in buildings.

We can use the same types of conductor for batteries that we find in **Chapter 3** in this book or sometimes we use flexible fine stranded cables from **Article 400: flexible cords and cables**.

In this case of the following answers, THWN is correct. THWN is a wire type that is often used in buildings. If we break it down, thermoplastic heat wet nylon (THWN), which means it can get to be 75C and wet. If it was THWN-2 then it could get to be 90C.

For battery series connections, we would use a larger THWN than for PV module interconnects, due to the higher currents.

37. Which of the following would be the least appropriate use of flexible fine-stranded cables?

a. **Using with screw terminal lugs**
b. Using with batteries
c. Using with compression connectors
d. Using with 4/0 wire

Flexible fine-stranded cables are often used in PV system battery banks where there are large conductors due to higher currents. The reason people use the fine-stranded cables is because they are easier to bend and work with.

One of the problems that people have with these cables is making a good connection to a lug. If an improper lug is used, then there will not be a good connection with the wire and when there is a bad connection there is resistance. Resistance and current cause heat and voltage drop. The heat can melt things and cause a fire. When there are hot battery cells, it can also cause faster loss of electrolyte due to the higher temperature.

The best way of overcoming this problem is to not use fine-stranded cables, to use proper compression lugs and to **beware of using typical screw terminal lugs**. The fine-stranded cables when tightened with a screw terminal, may seem tight for a while, but then can loosen up over time as the fine strands move over time.

There is no rule against using fine-stranded cables with batteries and when the conductors get as big as 4/0 then fine-stranded cables are often recommended.

38. Which would be a wiring method appropriate for on a roof in sunlight and in an attic for PV source circuits?

- **a.** Rigid non-metallic conduit
- **b.** PVC
- **c.** PV wire
- **d.** **EMT**

In Europe and much of the world, it is common to run PV wire from the PV all of the way to the inverter, however, **the NEC only allows PV source and PV output circuits inside a building that are in a metal raceway of in metal clad (MC) cable**.

After a readily accessible dc disconnect, it is allowable to have PV source or PV output circuits outside of a metal raceway, but is not common practice. The reason we like to keep the circuits inside of metal is because there may be a short circuit that would cause arcing, which would hopefully be contained inside of the metal raceway. Also metal raceways are good at detecting ground faults.

The only wiring method out of the questions above that is able to be used inside and is a **metal raceway is EMT**.

39. The UL label on an inverter says that the inverter can be used for ungrounded arrays. In which situation can it also be used for grounded arrays?

- **a.** If it will operate ungrounded, it can be used grounded, which is safer.
- **b.** If the manufacturers' instructions say it can be used for bipolar arrays.
- **c.** **If the UL label also said it can be used for grounded arrays.**
- **d.** If GFCI is installed.

If a typical non-isolated (aka ungrounded) inverter is grounded, it will indicate a ground fault and will not work. Technically, if an inverter were listed for ungrounded arrays, then it would also have to be listed for grounded arrays in order for it to be used. Remember to make sure your equipment was listed and tested for its intended purpose.

40. The dimensions of a 250W PV module are 1670mm × 990mm and there is a 5kW system. The wind uplift forces are 23lbs per square foot. The system is mounted in four rows with eight standoffs per row supporting the array. What would be the amount of uplift force per standoff?

- **a.** 120lbs
- **b.** 232lbs
- **c.** **256lbs**
- **d.** 44 newtons

First, we are going to determine how many 250W solar modules there are in 5kW:

250W/1000 watts per kW = 0.25kW per module
5kW/0.25kW per module = 20 solar modules

The next step is to figure out the area of the array in square feet, so we can start by converting the area of the dimensions of the modules from mm to feet:

3.28 feet = 1 meter

If we do not remember this conversion factor, we can look in the NEC and see that a lot of dimensions are given in both feet and meters and we can figure out the conversion factor that was used in the NEC.

Since the dimensions of the modules are given in mm, it is easy enough to convert mm to meters:

1670mm/1000mm per meter = 1.67m
990mm/1000mm per meter = 0.99m

Then convert to feet:

1.67m × 3.28 feet per meter = 5.48 feet
0.99m × 3.28 feet per meter = 3.25 feet

Then we multiply length by width to get area of a module:

5.48 feet × 3.25 feet = 17.8 square feet per module

Now we determine the area of the array (20 modules):

17.8 square feet per module × 20 modules = 356 square feet array

Remembering that the wind uplift forces are 23lbs per square foot:

23lbs per square foot × 356 square feet = 8188lbs of uplift force on array

Now we need to calculate the number of standoffs:

Four rows × eight standoffs per row = 32 standoffs

And for the final calculation, determining pounds of uplift per standoff:

8188lbs/32 standoffs = 256lbs per standoff

> We can see that this calculation is many easy steps that are not too difficult when taken one at a time, however, if you are nervous or not well rested, all it will take is one small mistake to get it wrong. This is why it is important to get a good night's sleep and to be prepared for an exam, in order to have the best outcome.

For Questions 41 through 45, use the following manufacturers' data:

Table 11.4 PV module data

Voc	37.8V
Vmp	31.1V
Isc	8.28A
Imp	8.05A
Temperature coefficient Voc	-0.30%/C
Temperature coefficient Vmp	-0.45%/C

Table 11.5 Inverter data

Power	5kW
Max input voltage	600V
MPPT voltage range	280–480V
Number of MPPT inputs	1

Table 11.6 Temperature data

Low design temperature	-13C
High cell design temperature	69C

41. What is the maximum number of modules in series?

- **a.** **14**
- **b.** 12
- **c.** 13
- **d.** 22

There are four things we need to determine the maximum number of modules in series:

1. Low design temperature = –13C
2. Temperature coefficient of Voc = -0.30%/C
3. Voc = 37.8V
4. Max inverter input voltage = 600V

Step 1: Determine delta T (difference in temperature)

–13C – 25C = –38C

Step 2: Determine percent change in voltage

–38C × –0.30%/C = 11.4% increase in Voc when cold

Step 3: Calculate increase in voltage: Method 1 of calculating increased voltage

0.114 × 37.8V = 4.31V increase at cold temperature
4.31V + 37.8V = 42.1 Voc per module cold

Method 2 of calculating increased voltage (shortcut)

1.114 × 37.8 = 42.1 Voc per module cold

Step 4: Determine maximum number of modules in series

600V/42.1V = 14.25

Round-down for max. in series calculation (rounding-up would go over voltage)

14 in series maximum

42. What would be the minimum number of modules that you would put in series, so that the inverter works efficiently when it is hot?

a. 7
b. **12**
c. 8
d. 10

This calculation is similar to the calculation done for the maximum in series, except we use the Vmp instead of the Voc and we will round-up at the end instead of rounding down. Also, we use the temperature coefficient of Vmp and the high cell temperature. This calculation is for efficiency and not done for safety. The worst-case scenario here would be having a PV system that does not work in the hot weather.

There are **four things that we need** to determine the minimum number of modules in series:

1. **High solar cell temperature = 69C**
2. **Temperature coefficient of Vmp = –0.45%/C**
3. **Vmp = 31.1V**
4. **Minimum inverter input MPPT voltage = 280V**

Step 1: Determine delta T (difference in temperature)

69C – 25C = 44C

Step 2: Determine percent change in voltage

44C × –0.45%/C = 19.8% decrease in Vmp when hot

Step 3: Calculate decrease in voltage: Method 1 of calculating decreased voltage

0.198 × 31.1 Vmp = 6.16V decrease at hot temperature
31.1 Vmp – 6.16V decrease = 24.9V per module hot

Method 2 of calculating increased voltage (shortcut)

19.8% decrease/100% = 0.198
1 – 0.198 = 0.802 (means we keep 80%)
0.802 × 31.1 Vmp = 24.9V per module hot

Step 4: Determine minimum number of modules in series

280V/24.9V = 11.2 modules in series

Round-up for minutes in series calculation (rounding down would go under voltage and inverter would shut off).

12 in series maximum.

43. If 20 modules will fit on the south-facing roof and 12 would fit on the southeast-facing roof, what would be the maximum amount of PV that you can use on this job with one inverter and these modules?

a. 3.8kW
b. 4kW
c. 5kW
d. **6kW**

In solving Questions 42 and 43, we determined that the PV source circuit (string) lengths for this job have to be between 12 and 14 in series. Since this inverter has one MPPT input, then all of the source circuits have to be the same length. Since only 12 would fit on the southeast roof, we would have to also put 12 on the south roof, even though more would fit, because all PV source circuits have to be the same length.

Therefore, the most we can put would be 12 on the southeast roof and 12 on the south-facing roof for a total of 24 PV modules.

To determine the power of each module, multiply operating current × operating voltage:

Vmp × Imp = power
31.1V × 8.05A = 250W

The module is a 250W module.

> Do not multiply Isc or Voc for power, since Isc and Voc do not produce any power.

If we have 24 modules that are 250W each, then:

24 modules × 250W = 6000W
6000W/1000W per kW = 6kW of PV

6kW is the correct answer here, but let's have a discussion about putting 6kW on a 5kW inverter.

> It is acceptable to put more PV on an inverter than the output rating of the inverter. This is done often. Here are some reasons why we can put more PV than stated on the inverter:
>
> 1. If the PV could make more than the inverter output, it would not damage the inverter and the inverter would clip power (clipping means the inverter would put out 5kW in this case even if the array had the potential to put out more using a larger inverter).
> 2. There are arrays facing different directions in Question 43. This means that both arrays will not be producing peak power at the same time.
> 3. When the sun is out it will get hot and the array will lose power.
> 4. There can be more options for connecting the system on the ac side if the inverter is rated for less power.
> 5. Smaller inverters and associated installation costs are less.
> 6. There are many derating factors that will decrease the dc power before it will get to the inverter.
> 7. Some large companies use 1.4:1 PV to inverter ratios.

44. In the system design in Question 43 on the label at the dc disconnect, what number will we put for maximum system voltage?

a. 600V
b. 480V

c. **505V**
d. 589V

In Section 690.17(A),, it says that if temperature coefficients are given, that we have to use them to calculate maximum system voltage. It is a common mistake by solar installers to put the maximum voltage of the inverter, which in this case is 600V. Instead we need to do the calculation for determining the Voc for a module for cold temperatures and then multiply that by the number of modules in series.

In this case, the system designed in Question 43 has 12 modules in series.

In Question 41, we calculated the cold temperature module Voc to be **42.1V per module.**

Therefore the maximum system voltage is:

12 in series × 42.1V per module = 505V

45. For the system in Question 43, what is the smallest allowable PV source circuit conductor size for PV wire jumpers behind the array? Assume 33C ASHRAE hot design temperature, 75C screw terminals, and 15A fuses in the combiner. Assume copper conductors.

a. 10 AWG
b. 12 AWG
c. **14 AWG**
d. 16 AWG

Here is the information that we will need to solve this problem:

1. Module Isc = 8.28A
2. NEC Table 310.15(B)(17): ampacities of conductors in free air
3. NEC Table 310.15(B)(2)(a): ambient temp correction factors

Here are the steps:

1. In NEC 240.4, note smallest conductor based on fuse size. Conductor must be at least this size.
2. Multiply Isc x 1.56
   a. Choose adequate conductor from Table 310.15(B)(17), use 75C column because of 75C terminals. Conductor must be at least this size.

3. Multiply Isc x 1.25

 a. Derate for heat with Table 310.15(B)(2)(a) (divide by derating number).
 b. Chose adequate conductor from Table 310.15(B)(17), use 90C column, conductor must be at least this size.

There are three different things that we do to size the wire and we pick a conductor that is large enough to satisfy all three ways of doing things.

Now we will break it down for a PV source circuit with the module given in the preceding table. Since current does not increase with a series connection, the current of a module is the same as the current of a PV source circuit.

1. In Article 240: overcurrent protection at 240.4 (D)(3) states that the smallest conductor that can be used with a 15A overcurrent protection device is a **14AWG** copper wire.
2. Isc x 1.56 = 8.28 x 1.56 = 12.9A

Table 11.7 From Table 310.15(B)(17)

Size AWG	75C	90C
18AWG	-	18A
16AWG	-	24A
14AWG	**30A**	35A
12AWG	35A	40A
10AWG	50A	55A

3. Isc x 1.25 = 8.28A x 1.25 = 10.4A

 a. We then have to derate for the ambient temperature of 33C, which we will do by looking at Table 310.15(B)(2)(a) in the 90C column, since PV wire is 90C rated wire.

Table 11.8 From Table 310.15(B)(2)(a)

Ambient Temp	75C conductor	90C conductor
26–30C	1.00	1.00
31–35C	0.94	0.96
36–40C	0.88	0.91

In this case, we will cross-reference a 33C ambient temperature with a 90C rated conductor and will get a 0.96 derating factor.

Since we will need a larger wire due to higher temperatures, then we need to make the current higher number and will have to divide 10.4A by 0.96 to make the number larger.

10.44A/0.96 = 10.8A

We will once again look at Table 310.15(B)(17), this time in the 90C column and see that here the **smallest acceptable wire is an 18 AWG** wire, which can carry 18A.

The three different steps (pathways) gave us three answers, two the same and one different: 14 AWG, 14 AWG, and 18 AWG wires. We will have to choose wire that will be the **largest wire of the three steps, which is a 14 AWG wire.**

Strategy: Think of what you are doing and what you want your number to do. If your number should get larger, then you should divide by numbers smaller than one and multiply by numbers greater than one. If you want your number to get smaller, you should multiply by numbers less than one and divide by numbers greater than one. Always think about what you are doing and what your numbers should be doing. People who get lost in formulas without thinking about what they are doing often make mistakes.

Oftentimes, we are surprised at how we could have used a conductor that is smaller than we thought. In this case, we were allowed to use a 14 AWG wire. Almost nobody in the industry uses 14 AWG conductors for PV source circuits. Here are a few reasons why we usually use a larger wire.

1. Voltage drop: If we use a wire that is smaller, there is more power loss. It is not worth installing a PV system and then skimping on wire in order to save a few dollars in the short run and losing hundreds or thousands of dollars' worth of power in the long run.
2. It is easier to stock one type of wire on the truck.
3. The price is not much different in many cases.
4. Larger wires are stronger.

You may notice on an exam that you are able to use a small wire.

In Question 45, if we did not have fuses (which is often the case with 2 PV source circuits or less) and if we had no 75C terminals, we could see how it would be possible to use an 18 AWG wire. It is not something that we would do, but it would be code compliant.

46. There is a 33 module 10kW ground mounted system with three rows of crystalline PV mounted in portrait. The modules are 72 cells with cells in the typical 6 × 12 arrangement. On a sunny December 18 at 2p.m., one out of three rows of modules is having one row of cells shaded. What would be the expected power loss from the shading assuming that all of the modules are connected to a single inverter with a single MPPT?

- **a.** 4% loss
- **b.** **33% loss**
- **c.** 50% loss
- **d.** 90% loss

In conventional crystalline silicon solar modules, there are bypass diodes that are arranged so that when there is shading, that the shaded cells and the other cells that are in the same group of the shaded cells are bypassed. Since the typical module is arranged into three groups of cells and if cells from all three groups are shaded, then the entire module will be bypassed.

When the modules are mounted in portrait and an entire row of cells along the short edge of a module is shaded in a sunny day, all of the bypass diodes in the module will skip the entire module with the shaded cells.

Figure 11.9 Bypass diodes, courtesy of Solmetric

PV Module with Bypass Diodes

Current can bypass groups of shaded cells and sacrifice voltage to get through.
Typical PV has 3 sections and will skip 1, 2 or all 3 at once.

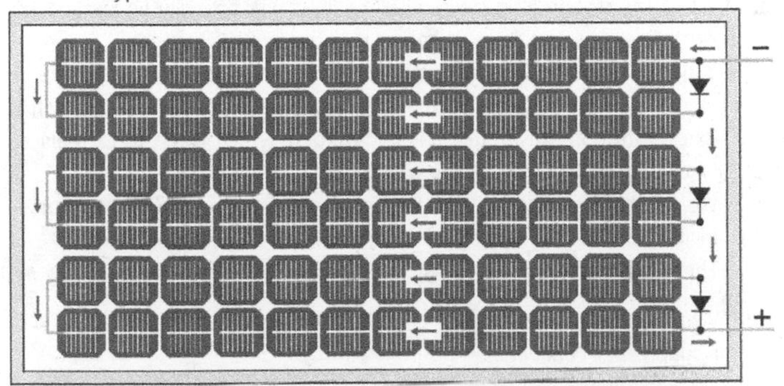

Shading of the long edge is more desirable than shading of the short edge due to the arrangement of the diodes. Shading a row on the short edge will take out all 3 sections

If the short edges of one-third of the PV modules are shaded, then the power output will decrease by one-third.

47. In a grid-tied system with battery backup, which of the following would be the most important attribute?

- **a.** Having a charge controller that will power diversion loads when the system is in interactive mode.
- **b.** Having a system that will equalize sealed lead-acid batteries.
- **c.** Using a multimodal charge controller with an automatic transfer switch.
- **d.** **Having an inverter that will power all of the backed up loads.**

The backed up loads with a grid-tied battery backup system are the loads that will operate when the grid is down. It would be important to have an inverter that will power all of the backed up loads. Typical backed up loads are a refrigerator, lighting, outlets and communication devices. Typically, people do not backup air-conditioning or heating, because those loads would very quickly drain the batteries.

To rule out some of the other wrong answers:

a. When the system is in interactive mode, it will typically be feeding the extra power back to the grid, rather than a diversion load.

b. You do not equalize sealed batteries. If the batteries are sealed, then you cannot add electrolyte to them. When you equalize batteries, the electrolyte will turn into hydrogen gas and oxygen, which needs to be replaced by distilled water.

c. We do not see multimode charge controllers in the NEC. A multimode inverter is what can operate in interactive and standalone mode.

d. **The correct answer.**

48. Which of the following would be the most important feature to have with a supply-side connection?

- a. **Have a service-rated fusible disconnect.**
- **b.** Do not exceed the rating of the main breaker.
- **c.** Do not exceed 120% of the ampacity of the service conductors.
- **d.** Do not exceed the rating of the busbar.

With a supply-side connection, you can feed as much power as the service can handle, which is a lot more than you would probably want to connect.

More specifically, in 705.12(A), it says that the sum of the ratings of overcurrent devices connected to power production sources cannot exceed the rating of the service.

That means if the service is 400A, that you can have 400A of solar breakers on the supply-side of the main breaker.

The reason that we can connect so much solar is that the loads and the busbar of the building are already protected by the main breaker (main service disconnect).

Since the supply-side of the main breaker is not protected by any other overcurrent protection device (OCPD) and since the supply-side connection is as vulnerable as any main service disconnect, then it is important that is **protected by a service-rated disconnect, whether it be a service-rated breaker of a service-rated fusible disconnect**.

49. On a low slope roof ballasted PV system, there are rooftop combiners that can take up to 20 strings each. If the PV Isc is 8.25A and the Imp is 7.9A and the PV source circuits each have 12 modules in series, then how many strings can each combiner have if there is a fused disconnect on the PV output circuit that is 200A?

a. **15**
b. 24
c. 20
d. 17

To size a fuse on a PV source or output circuit, we multiply Isc × 1.56 and round-up to the next common fuse size. If the fuse size was 200A, then to determine the amount of current that the fuse can protect, we will have to first determine what the current of 1 PV source circuit multiplied by 1.56 is:

8.25A × 1.56 = 12.9A

Then we take that current and divide it into 200A:

200A/12.9A = 15.5

Since we cannot have half of a circuit, we will have to round-down, so the **maximum number of PV source circuits is 15**.

> 156% is a number that is commonly known by those in the PV industry. We have this number to account for extra currents when PV modules can operate beyond STC. Sometimes irradiance in the natural world is above 1000W per square meter, which means that current will be above what is written on the label. This 1.56 correction factor is only on PV source and output circuits and not inverter output circuits, most dc-to-dc converter circuits, or standalone battery to inverter circuits.
>
> It is also interesting to note that the dimensions of the typical solar cell are 156mm × 156mm.

In **Section 690.9: overcurrent protection,** we find that **690.9(B) says that the OCPD shall not be less than 125%** of the currents from 690.8(A); when we are talking about PV source and output circuits, the current will be found in **690.8(A)(1), where it defines maximum current as 125% of Isc**. When we have 125% twice we have 1.25 × 1.25 = 1.56.

50. You are installing PV in a new laboratory for quantum computers where the temperature in the room is kept as cold as that in a freezer. You have conduit running through the cold room. What must you do for this rigid metal conduit that you would not otherwise have to do in a warmer environment?

- **a.** Conduit fill calculations due to the absence of heat.
- **b.** **Fill raceway with material to prevent circulation.**
- **c.** Install a heating element into the conduit.
- **d.** Allow air to circulate through the conduit.

Section 300.7: raceways exposed to different temperatures includes language for **sealing conduit with a material to prevent airflow, which would lead to condensation.**

Just like when you have a glass of iced water and condensation will build up on the outside, where there is conduit with temperatures that are different on the inside versus the outside of the raceway, the warmer side will have a tendency to produce condensation.

Section 300.7(A) is for sealing conduit to prevent condensation and 300.7(B) is to allow for expansion fittings. You should know how to look up expansion fittings, so it is convenient to find both in 300.7.

In the index of the NEC, you can look up "**Sealing**" (for sealing a raceway due to temperature) to find 300.7.

51. You are installing residential solar systems in Phoenix, Arizona, and need to make sure that you do not put too few modules in series. The data from the PV module is: power 225W, Voc 48.5, Vmpp 41, Impp 5.5, Isc 5.9, temperature coefficient Voc –133 mV/K, coefficient Vmpp –156 mV/K. The inverter specifications are: maximum voltage 600V and MPPT operating voltage 240V to 480V. The low design temperature is –13C, the high ambient temperature is 44C and the high design cell temperature is 75C. What is the least amount of modules that you can put in series without sacrificing performance?

a. 11
b. **8**
c. 5
d. 7

As with real life, there is more information than we need here. It is recommended that you separate out the information you need first:

1.	High cell temperature	75C
2.	Temperature coefficient of Vmpp	−156 mV/K
3.	Module Vmpp	41V
4.	Inverter low voltage	240V

> Some differences with this problem that we are not always used to is that the coefficient is given in mV/K instead of %/C. First of all, a change in degrees K is the same as a change in degrees C. There are 1.8 degrees F in change per 1 degree change C, but K changes are the same as C changes. The difference is that zero K, which is absolute 0, starts at −273C. Nothing in the universe can be colder than absolute0. mV or millivolts are thousandths of a volt, so It Is three decimal places to the left of a volt. It is important to make sure that you use a lower case m for milli and an upper case M for mega; there is a billion millivolts in a megavolt.

Here are the steps to solve this problem:

1. Find the difference in temperature from STC:

75C − 25C = 50C

2. Using the temperature coefficient for Vmpp, determine the hot temperature voltage:

50C × −156mV/K = −7800mV less voltage

3. Convert to volts:

−7800mV/1000mV per volt = −7.8 volts lost

Subtract lost volts from Vmpp:

41V – 7.8V = 33.2V when hot

4. Determine how many modules in series to keep inverter above lower end voltage by dividing hot PV Vmpp into low inverter voltage:

240V inverter/33.2V per module = 7.2 modules in series

In this case, we are going to **round-up**, since rounding-down would make the inverter turn off on a hot day. This calculation is not for a safety issue. The worst case here is having too few modules in series and having the inverter turn off on a hot day.

The answer is rounded-up to **eight modules in series.**

52. There are 15 PV source circuits and an equipment grounding conductor on a rooftop in sunlight in a circular raceway, which is elevated 3" above the roof going from the array to a combiner 22 feet away in a location with a constant breeze keeping the conduit cool. The ASHRAE 2% average high design temperature is 35C. The PV source circuits are 12 AWG THWN-2 and the ground wire is 12 AWG THWN-2. The terminals in the combiner box are rated for 75C. What is the maximum Isc for modules used in this system?

a. **10.35A**
b. 7A
c. 16A
d. 24A

First, we will list the information that is relevant to answering this question:

1. 35C high design temperature
2. 15 circuits = 30 current-carrying conductors
3. 12 AWG THWN-2

Note: Conduit 3" above the roof was relevant for answering this question when using the 2014 NEC. Temperature adders for raceway on a roof in sunlight were removed and no longer have to be accounted for.

Logic of what we are going to do: We are going to take the current of the conductor and then do different things to it that will decrease the original ampacity of the conductor, which we will find in the NEC.

We will look at the different tables that we will use from the NEC for solving this problem.

Table 310.15(B)(16) for ampacity of the conductor under 30C conditions. This table is for conductors that are not in free air, which includes conductors in conduit.

From Table 310.15(B)(16), we will find the ampacity of the conductor:

Table 11.9 From Table 310.15(B)(16) Allowable ampacity for conductors in conduit at 30C

Size of wire	75C rated conductor	90C rated conductor
14AWG	20A	25A
12AWG	**25A**	**30A**
10AWG	35A	40A

For determining the ampacity of a conductor for PV source and PV output circuits, there are two different calculations that we have to do and then we choose the larger conductor from the two separate calculations.

I like to call the the pathways:

1. 1.56 and terminal temperature limits
2. 1.25 and conditions of use (additionally we have to check that the conductor is big enough for the fuse using 240.4)

1.56 and terminal temperature limit step

I usually do the 1.56 and terminal temperature limits first, just because it is quick and easy. In the case of this problem, we have to work backwards, since we are solving for the Isc of the modules.

Since we have terminal temperature limits of 75C we are going to use **25A** for our first calculation, which corresponds to 12 AWG wire in the 75C column (note that in the NEC THWN-2 is in the 90C column, however we are still using the 75C column, since the terminals are not allowed to be above 75C).

$$25A/1.56 = 16A$$

Notice how we divide by numbers greater than one to make the number smaller. When we see the ampacity of the table, we want to make the number smaller to be safer.

In the 1.56 and terminals step, the greatest **Isc we can use is 16A**, but we are not done, because we have to do the 1.25 and conditions of use step.

Conditions of use are different conditions that cause the conductor to operate at a hotter temperature. There are two different conditions that we will be derating for:

1. Ambient hot design temperature is another condition of use that we will find in the NEC in Table 310.15(B)(2)(a). Often in the industry we use ASHRAE 2% average high design temperature that we can find at www.solarabcs.org.
2. More than three current carrying conductors in conduit is another factor that can add heat to the conductors. Too many conductors in conduit can heat the conduit and the wires.

All of these factors are going to decrease the ampacity of the conductors and all of these derating factors are less than one, so we will multiply the ampacity by these numbers in this situation.

If we are working the other direction and would like to start with the Isc of the module and work our way to the wire size, we would take the Isc of the module and do all of the steps in a way that makes the ampacity increase until we find our wire size in Table 310.15(B)(16) or in 310.15(B)(17), depending if the wire is in free air or not. Table 310.15(B)(17) is free air and 310.15(B)(16) is everything else.

1.25 and conditions of use step

This time we do not need to be concerned about the terminal temperature limits and we will start with the ampacity of 12AWG THWN-2 from the 90C column of Table 310.15(B)(16), which is **30A**.

We will then do a series of calculations, each making 30A smaller until we get our final ampacity.

First, we will make 30A smaller by a factor of 1.25 by dividing:

30A/1.25 = 24A

Next we have to derate for conditions of use.

35C ambient

Then we take the 35C temperature and look up the derating factor in Table 310.15(B)(2)(a). 35C falls between 56 and 60C. We can also see that 99% of the time and for our purposes that Table 690.31(A) is the same as Table 310.15(B)(2)(a). Our 90C rated wire here has a derating factor at 35C of 0.96.

Table 11.10 From 310.15(B)(2)(a) or 690.31(A) Derating for ambient temperatures other than 30C

Ambient temperature	75C conductor	90C conductor
26 to 30C	1.00	1.00
31 to 35C	0.94	**0.96**
36 to 40C	0.88	0.91

In this case, we are using the 90C column, since we are using THWN-2, which is a 90C rated conductor. We can easily turn the page and see that THWN-2 is a 90C conductor, since we see it written in the 90C column of Tables 310.15(B)(16) and 310.15(B)(17). Also, since it has a -2 after the THWN, that means that it is 90C rated.

Then we will make out ampacity decrease by a factor of 0.96 by multiplying by 0.96.

24A × 0.96 = 23A

The last derating factor for conditions of use that we are going to use is for more than three current carrying conductors in a raceway. In this case, we have 15 circuits and a ground. We do not count the ground and the 15 circuits each have two conductors, so we have a total of 30 current-carrying conductors.

Table 11.11 From Table 310.15(B)(3)(a) Derating for more than 3 current-carrying conductors in a raceway or cable

Number of conductors	Derating factor
4–6	80
7–9	70
10–20	50
21–30	45
31–41	40

The interesting thing about this table that is different from the other tables is that it gives us a percentage instead of a decimal derating factor. We can easily transpose the percent into a decimal and move the decimal two places to the left.

Since we have 30 current-carrying conductors, we have a derating factor of 45%:

45%/100% = 0.45 derating factor

We then need to derate our wire some more

$$23A \times 0.45 = 10.35A$$

We compare the 1.56 and conditions of use ampacity of 16A to the 1.25 and conditions of use derating ampacity of 10.35A and we determine that the most Isc that we can use from a PV module in this case is going to be the limit of 10.35A.

The answer is 10.35A.

Note: This is a very complicated wire sizing question. It gives you a good example of how difficult wire sizing can be. It is also not representative of the NABCEP exams, however. Exam questions are usually not this difficult. There are also good explanations of how to do wire sizing in my other book, *Photovoltaic and the National Electrical Code.*

53. You are running conduit from an array to a transition box in the roof and there are no terminal temperature limits given on the terminals, the transition box, or the installation manual for the transition box. At the transition box, the conductors will change from PV wire to THWN as the circuits go from the roof to the attic. Should you consider terminal temperature limits or what should you use for terminal temperature limits?

- **a.** Do not consider terminal temperature limits.
- **b.** **Use 60C terminal temperature limits.**
- **c.** Use 75C terminal temperature limits.
- **d.** Use 90C terminal temperature limits.

We can learn about terminal temperature limits in Section 110.14(C).

For connections that are less than 100A (or 14AWG to 1AWG) if we are not given the terminal temperature limits, we have to assume 60C terminals. (If over 100A or 1AWG, then we can assume at least 75C terminals.)

> Terminal temperature limits are usually 60C or 75C and many question why we use 90C wire at all. We still use the 90C column for derating for ambient temperature and when there are more than three current-carrying conductors in conduit. This goes for ac and dc.

54. What is the smallest copper equipment grounding conductor that can be used when running 10 source circuits in conduit with 10AWG THWN-2 wire 100 feet across a roof to a combiner box on the roof? The fuses in the combiner box are 15A and the combiner box terminals are rated for 75C.

- **a.** **14 AWG**
- **b.** 10 AWG
- **c.** 12 AWG
- **d.** 6 AWG

Equipment grounding conductor sizes are found in Table 250.122 and are based on fuse size.

Table 11.12 From Table 250.122 Minimum size of equipment grounding conductor (EGC)

Overcurrent device size	Copper EGC minimum size
15A	**14**
20A	12
60A	10
100A	8

From the table, we can see that the minimum size EGC would be a 14 AWG copper conductor.

It is acceptable to use a larger conductor and if the conductor is not inside of conduit, the AHJ may require it to be larger. In some locations the AHJ requires a 6 AWG when running bare copper in free air under arrays and in other locations the AHJ will require 10AWG bare copper in free air.

According to 690.45, if there were no fuses on a PV source circuit, we would use Isc x 1.56 in place of the OCPD in Table 250.122. The reason we can have no fuses on a PV source or output circuit is when there are one or two strings combined and not enough current for the fuse to open the circuit with one string backfeeding down another in a short circuit scenario.

55. A 10kW system with 250W modules in PV source circuits of 10 in series is producing about 75% of what was expected and you are sent to troubleshoot the system. First, you turn off the system and check each string individually at the combiner box, which is at the inverter. One of the strings measures 0V and all of the other strings measure within 2V of 351V. You then turn the system on and check the current of each string and notice that the string that measured 0V has more current than the other strings. All the modules are facing southwest at a tilt angle of 18°. What is the most likely problem?

- **a.** Ground fault
- **b.** **Short circuit**
- **c.** Uneven number of modules in different PV source circuits
- **d.** Polarity of one PV source circuit is reversed

When there is a short circuit on a PV source circuit, it will blow a fuse if there are enough PV source circuits to have current to blow the fuse. One or two strings will not have enough current to blow a fuse when there is a short circuit on a PV source circuit.

With the short circuit, we would measure 0V on the affected string and we would also measure short circuit current, which is greater than operating current. We would measure short circuit current when the system was off or when it is operating.

Figure 11.10 IV curve, courtesy of Solmetric

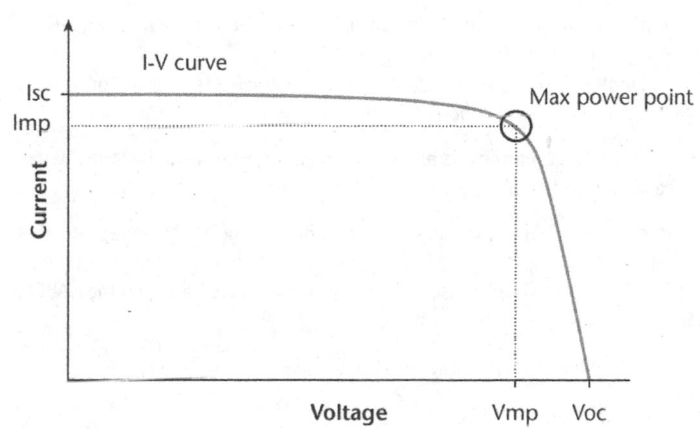

We can see in the IV curve in Figure 11.11 that at Isc, voltage is 0 and the current is at its highest point.

Short circuit current is always 0V and is always higher than maximum power current as we can see on an IV curve. The answer is we would see this scenario with a short circuit.

56. What is the correct OSHA practice?

- **a.** Have a safety meeting every day.
- **b.** **Tell workers about hazards.**
- **c.** Tell workers about their health and hospital plan.
- **d.** Provide earplugs to all employees.

OSHA is vast and the information can be overwhelming to learn. All of the OSHA material is free on the OSHA website. The question is which material is most important to study when preparing for the exam. Many students focus on the

PV portion of the exam and use logical reasoning, with experience and some studying of OSHA material to pass the exam.

OSHA stands for Occupational Safety and Health Administration.

Here are some good places on the OSHA website to study that will be an efficient use of your time:

- Top 10 hazards (starting with fall protection): https://www.osha.gov/Top_Ten_Standards.html
- Fall protection is most important in this industry: OSHA website for workers: https://www.osha.gov/workers.html

Here it states that **"employers must inform employees about hazards,"** which is the answer to the question.

There are plenty of links within these links. Remember to use your study time efficiently. If you are taking a NABCEP Exam, it is more than likely that you have already or will take OSHA classes, since it is a prerequisite for taking the NABCEP PV Installation Professional Exam.

57. Who removes the lockout tagout (LOTO) tags?

- **a.** The supervisor
- **b.** A specially designated person
- **c.** A licensed electrician
- **d.** **Whoever put the tags there**

Let us solve this problem with logic. Lockout tagout tags are placed typically on disconnects, so that we can turn something off without having to worry about someone else coming along and turning the switch on and electrocuting someone. It makes most sense that the person at risk for being shocked by someone turning the switch on is the person who should install and remove the lockout tagout tag.

> A funny way to remember this is you can think about putting the lockout tagout tag on before asking for a raise, so then they have to give you the raise in order to remove the tag. There are some exceptions to this rule, but in general, the person who puts the lockout tagout tag on will be the same person as the one who removes the lockout tagout tag.

Figure 11.11 Lockout tagout, courtesy of HH Barnum Company. www.hhbarnum.com/sites/ hhbarnum.onthevurge.net/files/05102012Brady-266.jpg

58. What is the biggest danger from seeing an arc from arm's distance?

- **a.** X-rays
- **b.** **UV rays**
- **c.** Infrared rays
- **d.** Shrapnel

UV rays are most likely to burn the retina, which is where your vision cells are. UV rays, which are what causes sunburn, can be focused by the lens of your eye onto the retina and cause damage. This can be referred to as arc eye and is related to snow blindness.

59. Of the following, what is the safest thing to do first when checking wires on a roof after a ground fault has been indicated?

- **a.** Remove fuses from combiner
- **b.** Turn on inverter
- **c.** **Inspect wires on the roof**
- **d.** Measure voltage

The least invasive thing is the best thing to do first. In this case, **inspecting wires is the least invasive** and involves the least amount of danger. The wires can even be inspected somewhat before climbing on the roof. Next, we could check the voltage and then remove fuses from the combiner, while wearing proper fall protection.

60. What is the safest test of the following?

- **a.** IV curve testing
- **b.** Measuring current through the meter
- **c.** **Measuring current with a clamp amp meter**
- **d.** Measuring voltage

Measuring current with a clamp is relatively safe, since you do not have to touch the wire being measured. IV curve testing and measuring current through the meter will have currents going through the meter of IV curve tester. Measuring voltage is relatively safe, however, you are still touching energized surfaces with your meter, which can be dangerous if you slip.

61. What is the device that is used for testing insulation of a conductor?

- **a.** **Megohmmeter**
- **b.** Refractometer
- **c.** Pyrometer
- **d.** Pyrheliometer

Here are the definitions of the devices above:

a.	**Megohmmeter**	**Used for testing the insulation of a conductor. Often called a Megger Tester, which is a brand name.**
b.	Refractometer	Tests the quality of battery acid in lead-acid batteries.
c.	Pyrometer	For measuring global solar radiation.
d.	Pyrheliometer	Measures direct beam solar radiation.

It is a good idea to remember the definitions of all of these meters above.

62. If the dc disconnect is not near the main service disconnect at the service entrance, what should be done according to the NEC?

- **a.** Install another dc disconnect at the service entrance. Under no circumstances can the dc disconnect not be at the main service disconnect at the service entrance.
- **b.** Include the location of all disconnects in the as-built plans that you will submit to the building department.
- **c.** **Have a sign at the disconnects in different locations that indicates where the other disconnects are located.**
- **d.** At the utility's discretion, you should put a sign at the main service entrance indicating where the dc disconnect is located.

There are many references to plaques and directories spread out through the NEC for PV:

- **690.4(D): multiple PV systems.** Multiple PV systems shall be permitted to be installed on a single building or structure. Where the PV systems are remotely located from each other, a directory in accordance with 705.10 shall be provided at each PV system disconnecting means.
- **705.10: directory.** A permanent plaque or directory denoting the location of all electric power source disconnecting means on or in the premises shall be installed at each service equipment location and at the location(s) of system disconnect(s) for all electric power production sources capable of being interconnected.

To sum this up and simplify: If all of the PV disconnects and the main service disconnect are all **at the same location, you do not need a plaque or directory**. If they are in **different locations, then at each of those locations, you do need a plaque** or directory showing where the other disconnects are located.

If we think about this, you would never have one plaque or directory, because if everything was not located in the same location, you would need at least two signs showing where the other disconnects are at two or more different places.

Therefore the correct answer to the question is "to have a sign at the disconnects at different locations that shows where the other disconnects are located."

This is one of those cases where it would be nice if the NEC had it all in the same location to avoid paper cuts.

63. You are connecting three different 5kW inverters to a subpanel on a residential service. Each inverter is connected to a 30A breaker on the subpanel. What is the smallest breaker that you could use for the breaker at the main service panel that is feeding the MLO (main lug only) subpanel?

a. 80A
b. 90A
c. 100A
d. 75A

First, we will determine the current that is coming out of the inverter and since it is a residential service, we are dealing with 240Vac:

Watts = volts × amps
Watts/volts = amps
5000W inverter/240Vac = 20.8A inverter output

Then we will multiply by 1.25 for continuous current and to size the circuit breaker:

20.8A × 1.25 = 26A

We have to round-up to the next common size of circuit breaker, so the test is right in using a 30A circuit breaker for the individual inverters.

To determine the smallest allowable breaker for the feeder, we will multiply 26A by three inverters. (Note: we did not use the 30A breaker size for this calculation.)

26A × three inverters = 78A

Rounding-up the next common size breaker gives us **80A, which is the correct answer.**

64. Using the information for Question 63 above, what would be the main breaker and busbar combination of the panelboard that would work best for a load-side connection?

a. 200A main breaker, 225A busbar
b. 400A center-fed main breaker, 450A busbar
c. **400A main breaker, 400A busbar**
d. 175A main breaker, 200A busbar

Let's see what the different options will let us connect on a load-side connection:

a. **200A main breaker, 225A busbar**

120% rule applies
busbar × 1.2 = 240A
240A – 200A main = 40A
40A/1.25 = 32A
Largest inverters here could only equal 32A
32A × 240V = 7680W inverter
definitely cannot connect three 5kW inverters here

b. **400A center-fed main breaker, 450A busbar**

Busbar is center fed here, so 120% rule does not apply.

The rule here is that the inverter current × 1.25 + main breaker cannot exceed the busbar rating according to 705.12(D)(2)(3)(a). This way you can put the solar breaker any place on the busbar, not just the opposite end from the main breaker, like with the 120% rule:

450A busbar – 400A main = 50A
50A/1.25 = 40A inverter(s)
40A × 240Vac = 9600W max. inverter(s)

Three 5kW inverters is more than 9600W or 9.6kW, so this will not work either.

c. **400A main breaker, 400A busbar**

400A busbar × 1.2 = 480A
480A – 400A main = 80A
80A/1.25 = 64A inverter(s)

64A inverter(s) × 240Vac = 15,360W of inverters
15.36kW > 3 × 5kW inverters

In this scenario, we can have three 5kW inverters! (**c.** is the correct answer.)

Let's solve this again another way with a formula that can let you keep all of the numbers on your calculator and do the calculation quickly!

We have spent a lot of time explaining these calculations, but once you practice the method, you can answer this problem veryquickly.

(((Busbar × 1.2) – main)/1.25) × grid voltage
= max. inverter power

Since the inverse of 1.25 is 0.8 we can also write:

(((Busbar × 1.2) – main) × 0.8) × grid voltage
= max. inverter power
(((400A × 1.2) – 400A) × 0.8) × 240Vac = 15.36kW

The calculator buttons we push are:

400 × 1.2 = – 400 = × .8 = × 240 =
and we get 15,360W

Practice this and you can get an answer in 30 seconds or under.

d. 175A main breaker, 200A busbar

(((Busbar × 1.2) – main) × 0.8) × grid voltage
= max. inverter power
(((200A × 1.2) – 175A) × 0.8) × 240V = 12.48kW

This is less than 15kW, but we would expect that, since we have already found the right answer. Thinking like a test taker can score extra points!

65. You are converting a grid-tied PV system to an ac coupled grid-tied battery backup PV system. Which of the following must you do?

- **a.** Use 50Vdc or higher rated disconnects for 48V battery circuits.
- **b.** Make provisions to equalize batteries if using sealed valve regulated lead-acid batteries. There shall be a schedule for equalization kept at the site of the battery bank.

c. **Have a sign that indicates grounded conductor, polarity, max short circuit current, and date calculations were performed.**

d. You must backup all of the circuits in the building. Selective loads are not allowed due to firefighters not knowing which loads are backed up.

Article 690, Part VI: marking is a good place to start looking for labeling requirements, also be aware that there are many places that we can look for PV and batteries in the NEC, such as **Article 480: torage batteries** or **690.55: PV systems connected to energy storage systems.**

Here are the two places where we find the answer to this question:

690.55: photovoltaic power systems employing energy storage systems (which is in Part VI: marking) says that PV connected to energy storage system should be marked with:

- Polarity

480.6: disconnecting means, 480.6(D): notification. Disconnect shall be labeled with:

- Nominal battery voltage
- Max. short circuit current
- Date calculation was performed

66. If a battery temperature sensor is not connected in New Jersey, what could be the problem with the batteries?

a. Undercharge in the summer and overcharge in the winter

b. Too much current in the winter

c. **Undercharge in the winter and overcharge in the summer**

d. Excessive current when it is hot

Batteries require more voltage to charge when it is cold. It is convenient that when it gets colder that PV makes more voltage and batteries require more voltage. If we did not have temperature compensation and were not increasing the voltage when it was cold, then the voltage would be low in the winter causing an undercharge and it would be high in the summer causing an overcharge. Another way to look at this is heat will speed up chemical reactions that happen in a battery, so the battery will require less voltage when it is hot.

If a battery overcharges in the summer, a symptom would be increased loss of electrolyte and a need to add distilled water. Another thing that can cause heat on a single cell is a bad connection at a battery terminal. The associated cell would require increased amounts of water when compared to the other battery cells.

The installation manual for the Rolls AGM battery recommends temperature compensation to be -4mV/C/cell. In a 48-volt system, which is 24 cells, our temperature coefficient would be - 96mV/C.

67. You are working on a project in Syria on a grid-tied 100kW project and a long piece of metal from an explosion has impaled a PV module. The array seems to be working fine and is feeding the grid, just like the day before. Of the following, which is likely the biggest problem?

- **a.** There is a ground fault.
- **b.** **The metal impaling the module would likely be energized and dangerous to touch.**
- **c.** There is an ac to dc short circuit.
- **d.** Lightning arrestors may have been activated.

If there were a ground fault, the inverter should have shut off, and since the inverter is still running, then there is probably not a ground fault. A short circuit would only remotely be possible. Besides the module being impaled, it would also have to have a conductor of the circuit electrically connected to something else that is energized at another voltage, which is not likely. There is no reason why lightning arrestors would be energized without lightning.

The most likely scenario here is that the metal that has impaled the module is energized and is at the voltage of the solar cells that it is touching. It would be dangerous to touch the metal.

68. The module interconnects should be:

- **a.** Anodized
- **b.** **Polarized**
- **c.** Galvanized
- **d.** Lubricated

The connectors between the modules are always polarized, which means that they are either positive or negative. There is no way to connect a typical PV module to another PV module with the connectors in any way but in series. To do parallel connections requires a combiner box or other equipment.

In **Article 690, Part IV: wiring methods** we find that **690.33(A): connectors** states that **connectors must be polarized.**

The connectors also have to click into place and should require a tool to open according to **690.33(C): type if over 30V**. Most modules operate over 30V.

69. Which of the following is the wrong way to do a supply-side connection?

- **a.** Split bolt onto service entrance conductors.
- **b.** Splicing into the conductors between the main breaker and the meter with a three-port lug.
- **c.** **Bolting conductors onto busbar on opposite side of main breaker from the meter.**
- **d.** Pull the meter before making the connection.

A supply-side connection is between the main service disconnect (main breaker) and the meter. **The wrong way to do a supply-side connection is anything that is not on the supply-side of the main breaker.** In the case of this question, bolting conductors onto the busbar on the opposite side of the main breaker from the main disconnect is the wrong way to make a supply-side connection.

Oftentimes, there is a main disconnect by the meter and a panelboard inside the house. Many people try and connect the solar system in between the main disconnect and the panelboard inside the house, which is also not a supply-side connection.

70. What are the best tools for checking performance of a system (commissioning)?

- **a.** **Irradiance meter, laser thermometer**
- **b.** Solmetric Suneye, inclinometer
- **c.** Tape measurer, camera, camera phone
- **d.** Solar Pathfinder, compass, level

To check the performance of a PV system, we want to see what the system is doing right now. What we would do is have an optimal performance estimate, as how the system would perform at STC and then derate the system for environmental conditions, which are temperature and irradiance. Most of the time temperature and irradiance are less than STC. The devices that we would use to measure temperature and irradiance are a **laser thermometer to get the cell temperature** of the module by measuring the backside of a solar cell. To get the irradiance of the array, we would have to have the **irradiance meter measure the irradiance** at the same tilt and azimuth angles as the solar array. Another name for an irradiance meter is a pyranometer.

When commissioning a PV system, we will take the irradiance measurement and use it as a derating factor. For instance, if we measured 870W per square meter, then our derating factor would be calculated as a fraction of STC, which is 1000W per square meter.

870W per square meter/1000W per square meter = 0.87

So our derating factor for irradiance would be 0.87, since we would be getting 87% of the irradiance that we would otherwise be getting under STC.

To derate for temperature, we need to use the temperature coefficient of power, which is usually found on the PV module datasheet. The temperature coefficient of power is different from the temperature coefficient of Voc and usually the same as the temperature coefficient of Vmp. A typical temperature coefficient of power would be about -0.48%/C.

We would get the cell temperature from the backs of solar cells with the laser thermometer. If the cell temperature is 55C, then we can calculate like this:

55C – 25C = 30C
30C × –0.48%/C = –14.4%

We would expect the performance to be derated for temperature by a 14.4% loss. To turn that into a decimal derating factor:

100% – 14.4% = 85.6%
85.6%/100% = 0.856

In the example here, we have derating for temperature to be 0.856 and derating for irradiance to be 0.87.

The total derating for environmental factors is then:

$$0.87 \times 0.856 = 0.745$$

On top of this, there would be other typical derating factors that we use for performance, such as inverter efficiency, wire loss, soiling, etc.

Solmetric Suneyes, Solar Pathfinders and other sun path calculators are not used for on the spot performance testing; they are used for predicting output of a PV system over a year.

Bonus practice exam questions

Chapter 12

30 questions

This is a bonus 30 question practice exam. This exam should take about 100 minutes if you would like to pace yourself as you would for a real 4 hour NABCEP exam.

1. An American solar module factory has a field where it is going to put a 100kW inverter. The field is 1000 feet from where they are going to do a supply-side connection at the main building of the factory. Exactly in between the field and the building there is a straw bale building that the CEO thought would be a good place to put the inverters. The inverters operate at about 400V dc and three-phase 480V ac. Where would be the best place to put the inverters to save money on copper?

   a. Straw bale building
   b. At the array
   c. Near the point of common coupling
   d. Between the array and the straw bale building

2. A new salesperson has just sold an array in a field that is 500 feet from a house. The typical operating voltage of the inverter is 380V dc and 240V ac. In between the house and the field is a tool shed that has no electricity. From an efficiency standpoint, where is the best place to put the inverter in order to reduce costs and increase performance?

   a. Tool shed
   b. Field
   c. Between the tool shed and field
   d. House

3. What is the time in which rapid shutdown shall take effect after being initiated?

a. 30 seconds
b. Immediately
c. 10 seconds
d. 5 minutes

4. The inverter company said the warranty of the inverter had been voided because it went over voltage 2 years before it broke. You want to dispute the inverter company and show it that it would have been impossible to go over voltage in the climate where the inverter was installed. There were 15 modules in series. The open circuit voltage of the modules is 35V and the temperature coefficient of open circuit voltage is –0.34%/C. What temperature would it have to get to in order for the system to have gone over the inverter and equipment maximum voltage of 600V dc?

a. -11C
b. -42C
c. -17C
d. -3C

5. On a ground-mounted 100kW system, PV source circuits on a bipolar system, the truck with the materials shows up and there are only 10AWG black-colored PV wire conductors on the truck, whereas you expected black, white, and red conductors for wiring behind the array tied to the racks. Can you use the black wire, and, if you can, why?

a. You must use white PV wire on the grounded conductors.
b. You can use the black wire if you mark it correctly.
c. 6 AWG and smaller conductors in this case cannot be marked on PV wire, only USE-2 wire for grounded systems.
d. As long as you separate the different monopole subarrays, you can use the black PV wire.

6. On sunny days, a utility interactive inverter has a history of repeatedly turning off for 5 or more minutes at a time. What is the most likely problem?

a. Inverter output circuit conductors are undersized.
b. String size is too long.

   c. Utility power factor is at 0.9.
   d. Low voltage ride-through.

7. On an ac-coupled multimode inverter, the dc connections are made to the

   a. PV source circuits
   b. PV output circuits
   c. Batteries
   d. MPPT charge controller

8. An inverter company is trying to void a warranty because it said there was an overvoltage. It is your job to determine how cold it must have been for the inverter to have experienced over 1000V dc on the dc input terminals of a ground-mounted 500kW array. The PV module STC-rated voltage is 44Voc and the temperature coefficient for the modules is –145 mV/C. Each source circuit has 20 PV modules. How cold must the PV have been for the overvoltage?

   a. –41.4C
   b. –16.4C
   c. –0.4C
   d. 2.4C

9. A 5kW 48V battery inverter is 92% efficient. The battery charging settings are set to work between 45V and 60V. What would be sufficient ampacity of the conductor between the battery and the inverter given these factors?

   a. 151A
   b. 204A
   c. 120A
   d. 221A

10. A residential PV system is put on a building that has a primary load of air-conditioning on when the sun is up. The customer is using a PV system that will offset 20% of his annual energy production. He is currently on a rate schedule of 15 cents per kWh. When solar is installed, he has an option for a time of use rate schedule, where daytime electricity rates are 20 cents per kWh and nighttime rates are 14 cents per kWh. Would it be beneficial to make the switch to time of use rates, and, if so, what is the reasoning?

- **a.** Yes, because when we are making energy it will be worth more.
- **b.** No, because nighttime rates are not that much of a saving.
- **c.** Yes, because it will offset the air-conditioning during peak times.
- **d.** No, because loads will be more than production.

11. You are hired to commission a 100kW STC PV system. Your irradiance meter reads 805W per square meter at the same tilt and azimuth as the array and the temperature sensor on the back of the PV module reads 42C. The temperature coefficient for power is –0.48%/C. Derating factors, including inverter efficiency, soiling, wire loss, PV module nameplate derating, etc., are 0.82. What do you expect the inverter power output to be?

- **a.** 50kW
- **b.** 60kW
- **c.** 70kW
- **d.** 80kW

12. Which are the correct calculations for the numbers written on the label at the dc disconnecting means at a utility interactive inverter fed by a PV source circuit?

- **a.** Isc × 1.25 × modules in parallel and temperature corrected open circuit voltage of array
- **b.** Imp × modules in parallel, Isc × 1.56 × modules in parallel, inverter maximum input voltage, and Vmp × modules in series
- **c.** STC power of array, Imp × modules in parallel, Isc × 1.56 × modules in parallel, temperature corrected Voc of array, Vmp of array
- **d.** Imp × 1.25 × modules in parallel, Isc × modules in parallel × 1.56, Voc × modules in series, Vmp × modules in parallel

13. A ground-mounted PV system with single rows of 72-cell PV modules mounted in portrait has 6" of thick snow covering the bottom row of cells of every module. If your system were usually making 50kW at noon on a normal day like this day when there is no snow, what would you expect would be the closest amount of power that the system would be making under these conditions at noon? Assume that very little light gets to the single row of cells and all other rows of cells get plenty of light and cold temperatures.

- **a.** 5kW
- **b.** 20kW

c. 45kW
d. 33kW

14. You are doing maintenance on an old PV system and the inverter needs to be replaced. You are replacing an inverter with a high-frequency transformer with an ungrounded inverter. The original inverter has the dc disconnect built into the inverter and the new inverter has a disconnect that is shipped with the inverter, yet is separable from the inverter. Which of the following is the correct procedure for replacing the inverter?

a. Turn off the ac and dc disconnects at the broken inverter, then test voltage on the inverter side of the ac and dc terminals in the inverter. Remove the inverter and replace with the new inverter.
b. Open inverter disconnects, disconnect a connector on the roof for each PV source circuit, test voltage, and replace the inverter.
c. Cover array with a tarp, then replace inverter.
d. Wait for nighttime, turn off disconnects, replace the inverter.

15. What is the slope of a 6:12 roof?

a. 45 degrees
b. 33 degrees
c. 20 degrees
d. 27 degrees

16. An existing feeder has two subpanels connected to it. The feeder is fed by a 150A breaker. One of the subpanels is protected at the subpanel by a 60A breaker and the other subpanel is protected at the subpanel by a 150A breaker. You are adding a 32A interactive inverter to the feeder. What would be the minimum ampacity of the conductor going from the main feeder to the 60A breaker protecting the subpanel that is 24 feet from the connection to the feeder?

a. 60A
b. 63A
c. 19A
d. 120A

17. Is it acceptable to have 100A of solar breakers feeding a 100A subpanel? If so, what must be done? If it is not acceptable, then why not?

a. It is not acceptable because it exceeds the 120% rule.
b. It is acceptable if there are no loads and there is a special label.
c. It is acceptable if it is an MLO subpanel only.
d. It is not acceptable because it will overload the busbar.

18. Which of the following scenarios would most likely NOT comply with rapid shutdown of PV systems on buildings for utility-interactive systems?

a. Microinverters on a hospital
b. Dc-to-dc converters that take two modules in series on and industrial rooftop
c. BIPV with no metal parts on a roof with no metal parts with a contactor disconnect inside the array
d. A ground mount with an inverter mounted on a residential wall with a contactor disconnect between the ground mount and the wall.

19. If a PV module has anodized aluminum frames and is installed with anodized aluminum rails, what is true about grounding or bonding of the modules?

a. The grounding from the array to ground must have a continuous or irreversibly spliced grounding conductor to earth.
b. If it is an ungrounded array, then you do not need to ground the array.
c. If you have a letter from the manufacturer saying you can use WEEB bonding washers and the bonding washers are UL listed, you CANNOT use them unless bonding washers are mentioned in the installation manual for the modules.
d. Copper equipment grounding conductors CANNOT be used in a system with aluminum PV module frames.

20. On a PV system, one of your workers said that he touched a white wire in a combiner box and got shocked. What is the most likely reason for this? The system was installed in 2007.

a. Ground fault
b. Positive to negative short
c. Copper equipment grounding conductor touching aluminum on array
d. The installer was hit by lightning

21. A house has a 200A panelboard and a 200A busbar with 220A of loads already in the panelboard. In which situation can you install a 40A solar backfed breaker on the load-side inside the panelboard?

a. If the solar breaker is marked line and load.
b. If there is another 40A backfed breaker on the other end of the center-fed panelboard.
c. If the busbar is center fed and there are no other solar breakers installed.
d. Only if there are no quad or dual breakers doubling up on breaker spots.

22. At least how deep must PVC conduit be buried in a trench across a dirt yard with PV source circuits in the conduit?

a. 12"
b. 18"
c. 3 feet
d. 6 feet

23. You can fit and sell 24 modules on west- and south-facing roofs. What would be the best combination for maximum production? Your single inverter has one MPPT.

a. 12 modules on the west and 12 on the south
b. 19 modules on the south and five on the west
c. 13 modules on the south and 11 on the west
d. 24 modules on the west and zero facing south

24. Which is the best type of ladder to use when working with electrical equipment and for getting onto a roof?

a. Aluminum ladder
b. Painted wooden ladder with no metal parts
c. Fiberglass ladder with aluminum steps
d. A-frame fiberglass ladder with fiberglass steps

25. Rigid non-metallic conduit is going 250 feet in a location where the low temperature is −12C and the high temperature is 38C. How many expansion joints must be used? Each expansion fitting can travel 4".

- **a.** 1
- **b.** 2
- **c.** 4
- **d.** 3

26. In a positively grounded, solidly grounded direct water pumping PV system, what colors should the conductors be?

- **a.** White positive, black negative, green ground
- **b.** Red positive, black negative, bare ground
- **c.** Black positive, white negative, green ground
- **d.** Red negative, black positive, bare ground

27. In a residential 240V ac split-phase service, what is the relationship of the conductors to each other?

- **a.** The white wire is 120V from ground.
- **b.** Line 1 and Line 2 are both 120V from ground.
- **c.** Line 1 is at ground voltage.
- **d.** Line 2 is at ground voltage.

28. What is the best configuration for connecting lead–acid batteries, assuming that all of the combinations below add up to the same amount of energy in the battery bank and that the battery bank is 48V nominal voltage?

- **a.** 24 batteries in series.
- **b.** Two sets of 12 batteries in series in two different circuits paralleled together.
- **c.** Four sets of six batteries in series in four different circuits paralleled together on large busbars.
- **d.** Eight sets of four batteries in series paralleled together.

29. What is the minimum size of an equipment grounding conductor in conduit for a PV source circuit of 12 modules in series with modules? There are no fuses or overcurrent protection devices on the source circuit. The following specifications of the modules are: Voc 44V, Vmp 35V, Isc 8.5A, Imp 7.8A.

- **a.** 14 AWG copper
- **b.** 12 AWG copper

c. 10 AWG copper
d. 6 AWG copper

30. A charge controller has a maximum input voltage of 150V and a minimum recommended operating voltage of 65V. The module you are using has the following specifications: Voc 44V, Vmp 35V, Isc 8.5A, Imp 7.8A, temperature coefficient for Voc = -0.34%/C, temperature coefficient for Vmp = -0.48%/C. The location has a high design cell temperature of 60C and a low design ambient temperature of -22C. What would be the possible source circuit configuration(s)?

a. Two or three in series
b. Only three in series
c. Three or four in series
d. Only two in series

Bonus practice exam questions

Chapter 13

With answers and explanations

Learning from going through these detailed explanations is one of the best ways to learn.

1. An American solar module factory has a field where it is going to put a 100kW inverter. The field is 1000 feet from where they are going to do a supply-side connection at the main building of the factory. Exactly in between the field and the building there is a straw bale building that the CEO thought would be a good place to put the inverters. The inverters operate at about 400V dc and three-phase 480V ac. Where would be the best place to put the inverters to save money on copper?

    a. Straw bale building
    b. At the array
    c. Near the point of common coupling
    d. Between the array and the straw bale building

In general, we say that the best conductors to go a long distance are the conductors that are the highest voltage. In this case, at first glance 480V is the higher voltage and it would be best to put the inverter at the array to reduce losses from voltage drop.

There is another reason why it is even more efficient to put the inverter by the array: with three-phase inverters and equipment, there is an extra benefit by the magnitude of the square root of 3, which is about 1.73. This would give us even more reason to put the inverter at the array. When doing the math for voltage drop and the resistance of the wire, if you have single phase, you will have to double the distance for the wire, since the distance of the wire is to and from.

With three phases, your efficiency will be better to the degree of the square root of 3 divided by 2, which is 0.866. A shortcut is to calculate voltage drop for single phase then multiply by 0.866 to get three-phase voltage drop.

2. A new salesperson has just sold an array in a field that is 500 feet from a house. The typical operating voltage of the inverter is 380V dc and 240V ac. In between the house and the field is a tool shed that has no electricity. From an efficiency standpoint, where is the best place to put the inverter in order to reduce costs and increase performance?

  a. Tool shed
  b. Field
  c. Between the tool shed and field
  d. **House**

In this case, the higher voltage should be the longest run possible and the lower voltage should be the shortest. With dc and ac, when it is not three phase, it is the same in comparison. In this case, **380V dc > 240V ac then the 380V dc should be optimized** and 240V ac reduced as much as possible. Therefore, in this case it would be best to have the inverter at the house.

When we talk about operating voltage, on the ac side it is 240V – in this case from the utility – but on the dc side it will change. The modules will make most of their power at a voltage close to, but lower than, Vmp due to heat causing voltage to be lower.

3. What is the time in which rapid shutdown shall take effect after being initiated?

  a. **30 seconds**
  b. Immediately
  c. 10 seconds
  d. 5 minutes

Rapid shutdown shall take place within 30 seconds. In the 2014 NEC, there was a provision for 10 seconds that was changed by a TIA (temporary interim amendment) changing mid-cycle the NEC to 30 seconds. The voltages that must go down in 30 seconds are 30V in 30 seconds for controlled conductors outside of the array boundary and 80V in 30 seconds within the array boundary. The array boundary is defined as being within 1 foot of the array.

4. The inverter company said the warranty of the inverter had been voided because it went over voltage 2 years before it broke. You want to dispute the inverter company and show it that it would have been impossible to go over voltage in the climate where the inverter was installed. There were 15 modules in series. The open circuit voltage of the modules is 35V and the temperature coefficient of open circuit voltage is -0.34%/C. What temperature would it have to get to in order for the system to have gone over the inverter and equipment maximum voltage of 600V dc?

a. -11C
b. -42C
c. **-17C**
d. -3C

In order for 15 modules to get to be 600V dc, each module must be 1/15 of 600V dc, which is calculated by:

600V/15 modules = 40V per module

If the modules were at first 35V, then the increase in voltage must have been:

40V – 35V = 5V increase in voltage

The percentage increase in voltage would be calculated by:

5V/35V = 0.1430
143 × 100% – 14.3% increase in voltage

If each degree decrease in temperature would increase the voltage by 0.34%, then to determine the change in temperature from STC we would do the following calculation:

14.3% divided by 0.34%/C = 42C change in temperature

Since the voltage was higher, then the temperature was colder and 42C colder than STC of 25C is:

25C – 42C = –17C

In order for that system to go over 600V, the solar cells would have had to be -17C when there was sunlight.

5. On a ground-mounted 100kW system, PV source circuits on a bipolar system, the truck with the materials shows up and there are only 10AWG black-colored PV wire conductors on the truck, whereas you expected black, white, and red conductors for wiring behind the array tied to the racks. Can you use the black wire, and, if you can, why?

a. You must use white PV wire on the grounded conductors.
b. **You can use the black wire if you mark it correctly.**
c. 6AWG and smaller conductors in this case cannot be marked on PV wire, only USE-2 wire for grounded systems.
d. As long as you separate the different monopole subarrays, you can use the black PV wire.

Bipolar PV systems, by definition, have grounded conductors on the dc side. With a bipolar system, there are two grounded arrays, one being positively grounded and the other negatively grounded.

In the NEC, there are rules about grounded conductors 6 AWG and smaller having to be usually white or gray, and there is a special provision for PV systems.

NEC 200.6(A)(6) says that single-conductor, sunlight-resistant, outdoor-rated cable used as a grounded conductor in a PV power system, as permitted in 690.31, shall be **identified** at the time of installation by **distinctive white markings** at all terminations.

This means that, for the grounded conductor of PV source circuits that are tied behind the racks, we can mark them white at all terminations.

The monopole subarrays should also be kept separated regardless of the color of the conductors.

The answer to this question would be the same for any PV system, not just bipolar systems, although ungrounded systems would not have a white-marked grounded conductor.

6. On sunny days, a utility interactive inverter has a history of repeatedly turning off for 5 or more minutes at a time. What is the most likely problem?

a. **Inverter output circuit conductors are undersized.**
b. String size is too long.
c. Utility power factor is at 0.9.
d. Low voltage ride-through.

Undersized conductors causes voltage drop and voltage drop causes the voltage to be lower where the power is going to than where the power is coming from. On an inverter output circuit, the power comes from the inverter to the point of interconnection. Having an undersized interactive inverter output circuit causes the voltage to rise at the inverter due to voltage drop on the conductor, since the voltage at the interconnection is, for the most part, set by the utility, the only way to account for voltage drop is to have a higher voltage at the inverter.

7. On an ac-coupled multimode inverter, the dc connections are made to the

- **a.** PV source circuits
- **b.** PV output circuits
- c. **Batteries**
- **d.** MPPT charge controller

Figure 13.1 AC-coupled PV system connections

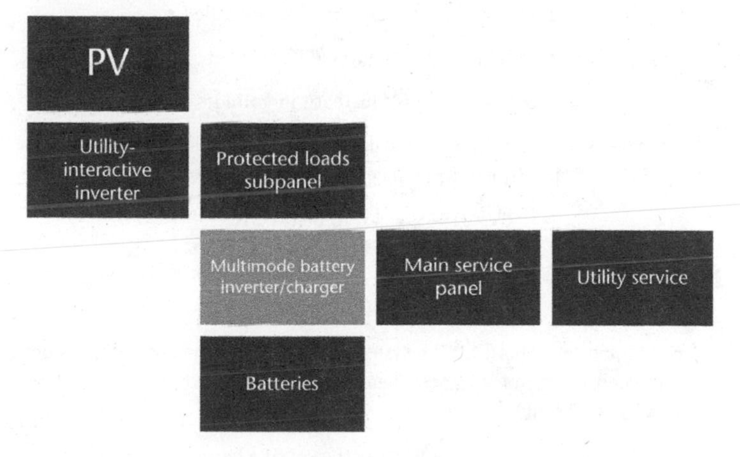

On an ac-coupled system, there are two different types of inverter, a multimode inverter that is connected to a battery bank and utility-interactive inverter(s) that are connected to a protected load subpanel. The utility is then connected to the multimode inverter. When the utility is disconnected from the multimode battery inverter, then the utility-interactive inverters will see the voltage from the multimode inverter, which is an ac voltage source, and then produce power when the voltage is in the correct range for the multimode inverter to work.

8. An inverter company is trying to void a warranty because it said there was an overvoltage. It is your job to determine how cold it must have been for the inverter to have experienced over 1000V dc on the dc input terminals of a ground-mounted 500kW array. The PV module STC-rated voltage is 44Voc and the temperature coefficient for the modules is –145 mV/C. Each source circuit has 20 PV modules. How cold must the PV have been for the overvoltage?

- **a.** -41.4C
- **b.** **-16.4C**
- **c.** -0.4C
- **d.** 2.4C

For 20 modules in series to reach 1000V dc, then each module must have reached:

1000V/20 modules = 50V per module

The increase in voltage for each module would have been:

50V – 44V = 6V increase in voltage

If each module increased 6V and each degree C increase will increase the modules 145 mV or 0.145V, then the change in temperature must have been:

6V/0.145V = 41.4C below STC

41.4C below STC of 25C is:

25C – 41.4C = –16.4C

It must have been **–16.4C** for the system to go over voltage and if the system were in Puerto Rico, for instance, then it would not have gone over voltage, since in Puerto Rico it never freezes.

9. A 5kW 48V battery inverter is 92% efficient. The battery charging settings are set to work between 45V and 60V. What would be sufficient ampacity of the conductor between the battery and the inverter given these factors?

- **a.** **151A**
- **b.** 204A
- **c.** 120A
- **d.** 221A

If an inverter can continuously put out 5kW and is 92% efficient, then the power coming into the inverter must be:

5kW/0.92 = 5.43kW (5430W) coming from the battery

The highest current coming from the battery would be at the lowest voltage setting, so the most current that can be coming from the battery would be:

$W = VI$ $I = W/V$ = **5430W/45V = 121A**

Since 121A could be coming from the battery to the inverter continuously, then the correction factor for continuous current is 1.25 so:

121A × 1.25 = 151A ampacity required

When sizing the circuit from the battery to the inverter there is no need for an extra 1.25 correction factor (1.25 × 1.25 = 1.56). Many solar designers get so used to using the 1.56 correction factor with dc that they accidentally use it for a battery to inverter circuit.

10. A residential PV system is put on a building that has a primary load of air-conditioning on when the sun is up. The customer is using a PV system that will offset 20% of his annual energy production. He is currently on a rate schedule of 15 cents per kWh. When solar is installed, he has an option for a time of use rate schedule, where daytime electricity rates are 20 cents per kWh and nighttime rates are 14 cents per kWh. Would it be beneficial to make the switch to time of use rates, and, if so. what is the reasoning?

a. Yes, because when we are making energy it will be worth more.
b. No, because nighttime rates are not that much of a saving.
c. Yes, because it will offset the air-conditioning during peak times.
d. **No, because loads will be more than production.**

If a building's primary load is when the sun is up and he is using more energy than he is producing, then switching to time of use rates will increase the electric bill. In extreme cases, people have bought solar and gone to a time of use rate schedule and have had their bill go up after installing solar.

In the case described, if most of the solar energy was made when the air-conditioning was on, then we would rarely be exporting energy. In that case, the energy would cost more than it would without a time of use rate schedule. There may be some savings at night, but, in this case, very little energy was used at night.

To take this to an extreme level, imagine if there were a large air-conditioned, day-use building that used $5000 worth of energy per month and it bought one solar module and switched to a time of use rate schedule. In this case, the energy produced by the one module would have little significance and the electric bill would go up dramatically.

Exaggerated learning concepts

It can be beneficial to think of exaggerations in order to understand concepts.

The following are examples:

1. If you connect 5kW of PV to a MW inverter, you will size the dc side of the inverter for 5kW and the ac side of the inverter for a MW.
2. If you connect a MW of PV to a 5kW inverter, you will size the ac side of the inverter for 5kW and the dc side of the inverter for a MW. (In this case you would "clip" the MW to 5kW all day long.)

   1 and 2 dramatically demonstrate that you do not relate dc and ac wire sizing.

3. You can short circuit the dc side of a MW at the inverter and never blow a fuse in a combiner.

   3 demonstrates that Isc is not much more than Imp.

11. You are hired to commission a 100kW STC PV system. Your irradiance meter reads 805W per square meter at the same tilt and azimuth as the array and the temperature sensor on the back of the PV module reads 42C. The temperature coefficient for power is –0.48%/C. Derating factors, including inverter efficiency, soiling, wire loss, PV module nameplate derating, etc., are 0.82. What do you expect the inverter power output to be?

a. 50kW
b. 60kW
c. 70kW
d. 80kW

The derating for the irradiance will be calculated by:

805W per square meter/1000W per square meter = 0.805 irradiance derating

The derating for power will be calculated by first determining the difference in temperature from STC, which is:

42C – 25C = 17C

Then we will have a loss of power that we can calculate by multiplying the temperature difference by the temperature coefficient of power:

17C × –0.48%/C = –8.16% loss of power due to heat

To determine a loss, we can subtract 8.16% from 100%:

100% – 8.16% = 91.8% is the power we keep

Turn 91.8% into a decimal derating factor:

91.8%/100% = 0.918

Now our derating factors are

- 0.918 for power due to heat
- 0.805 for irradiance below STC
- 0.82 for everything else

Now we derate our 100kW of PV:

100kW × 0.918 × 0.805 × 0.82 = 60.6kW ac output

We would expect the system to put out about **60.6kW ac.**

12. Which are the correct calculations for the numbers written on the label at the dc disconnecting means at a utility interactive inverter fed by a PV source circuit?

a. Isc × 1.25 × modules in parallel and temperature corrected open circuit voltage of array

b. Imp × modules in parallel, Isc × 1.56 × modules in parallel, inverter maximum input voltage, and Vmp × modules in series
c. STC power of array, Imp × modules in parallel, Isc × 1.56 × modules in parallel, temperature corrected Voc of array, Vmp of array
d. Imp × 1.25 × modules in parallel, Isc × modules in parallel × 1.56, Voc × modules in series, Vmp × modules in parallel

Section 690.53 describes what should be on the label; the harder part is describing what that means:

1. Rated maximum power point current is: Imp × modules in parallel.
2. Maximum system voltage is: temperature-corrected open circuit voltage of array.
3. Maximum rated output current of charge controller or dc-to-dc converter (if installed).

Therefore the correct answer is **a**.

In previous versions of the NEC (2014 and earlier), we were supposed to put the Vmp and Imp calculated values, which is no longer the case.

13. A ground-mounted PV system with single rows of 72-cell PV modules mounted in portrait has 6" of thick snow covering the bottom row of cells of every module. If your system were usually making 50kW at noon on a normal day like this day when there is no snow, what would you expect would be the closest amount of power that the system would be making under these conditions at noon? Assume that very little light gets to the single row of cells and all other rows of cells get plenty of light and cold temperatures.

a. **5kW**
b. 20kW
c. 45kW
d. 33kW

If a single row of cells is shaded on the short edge of a PV module, then all of the bypass diodes of the module would typically kick in. In this case, however, if every single bypass diode would kick in, then there would be no current to bypass. In this case, the amount of current flowing through the system would be equivalent to the weakest link, which is proportional to the amount of light getting to the cells behind the thick snow.

In this case, probably the closest answer would be 5kW. The next closest answer, which is 20kW, is 40% of what you would expect on a similar day when there is no snow covering the rows.

14. You are doing maintenance on an old PV system and the inverter needs to be replaced. You are replacing an inverter with a high-frequency transformer with an ungrounded inverter. The original inverter has the dc disconnect built into the inverter and the new inverter has a disconnect that is shipped with the inverter, yet is separable from the inverter. Which of the following is the correct procedure for replacing the inverter?

- **a.** Turn off the ac and dc disconnects at the broken inverter, then test voltage on the inverter side of the ac and dc terminals in the inverter. Remove the Inverter and replace with the new inverter.
- **b. Open inverter disconnects, disconnect a connector on the roof for each PV source circuit, test voltage, and replace the inverter.**
- **c.** Cover array with a tarp, then replace inverter.
- **d.** Wait for nighttime, turn off disconnects, replace the inverter.

Some older inverters had built-in disconnects, which was unfortunate for those replacing the inverter. When the disconnect was turned off (opened), and if the inverter was removed, the voltage from the array would still be making its way to the inverter, since the disconnect was built into the inverter. The only way to safely remove the inverter is to make sure that all of the PV source circuits are not connected. The connectors on the roof that are marked "do not open under load" should then be opened after the load break-rated dc disconnect inside the inverter is opened.

Before touching or working with conductors, you should always inspect your multimeter, test your meter, and check the conductors and terminals that you are about to work near before working with the conductors.

On an inverter, when the dc side of the inverter is disconnected from the array, there is still potential for shock due to capacitors in the inverter, so you should also wait for the capacitors to discharge, which typically takes 5 minutes. You should, of course, test the voltage.

Working with inverters at night or when covered with tarps is safer, although it can still be dangerous. People have been shocked by PV arrays at night, since voltage is present even at low light levels . Nighttime light can be from the moon, streetlights, lightning flashes, and other lights. Also, most, if not all, PV arrays under tarps still experience light and voltage.

The correct answer is **b.**, since we have to disconnect the individual PV source circuits, which is typically done on the roof or within the array. Voltage always has to be tested, since the wrong conductors could have been opened.

The same procedure should be done when replacing a dc disconnect, which is commonly done when replacing inverters.

15. What is the slope of a 6:12 roof?

- **a.** 45 degrees
- **b.** 33 degrees
- **c.** 20 degrees
- **d.** **27 degrees**

Roofers memorize roof slopes and here is how to calculate the slopes:

Rise/run = inverse tangent on calculator

- 3/12 Inv Tan = 14 degrees slope
- 4/12 Inv Tan = 18 degrees slope
- 5/12 Inv Tan = 23 degrees slope
- 6/12 Inv Tan = 27 degrees slope

You can memorize the above slopes or, better yet, just learn the calculator keystrokes.

16. An existing feeder has two subpanels connected to it. The feeder is fed by a 150A breaker. One of the subpanels is protected at the subpanel by a 60A breaker and the other subpanel is protected at the subpanel by a 150A breaker. You are adding a 32A interactive inverter to the feeder. What would be the minimum ampacity of the conductor going from the main feeder to the 60A breaker protecting the subpanel that is 24 feet from the connection to the feeder?

a. 60A
b. **63A**
c. 19A
d. 120A

According to the 25 feet tap rule, we add the feeder supply breaker, which is 150A, to 125% of the inverter current, and then take one-third of the resulting current to size the tap conductor.

The inverter current is **32A**.

125% of the inverter current is 32A × 1.25 = **40A**.

Adding 125% of the inverter current to the supply breaker is 40A + 150A = **190A**.

One-third of 190A is 190A/3 = **63A**.

Therefore, the smallest the tap conductor can be is **63A**.

(Let it be known that there are no rumors of this material ever being covered on a NABCEP exam ... yet.)

17. Is it acceptable to have 100A of solar breakers feeding a 100A subpanel? If so, what must be done? If it is not acceptable, then why not?

a. It is not acceptable because it exceeds the 120% rule.
b. **It is acceptable if there are no loads and there is a special label.**
c. It is acceptable if it is an MLO subpanel only.
d. It is not acceptable because it will overload the busbar.

According to 705.12(B)(2)(3)(c), which was added to the NEC in 2014, we can have as many overcurrent protection devices as we want on the busbar, as long as they do not exceed the rating of the busbar. In this calculation, we do not include the supply-side main busbar OCPD. This means that if we have a 100A busbar, we could add 100A of solar, as long as we did not add any other breakers ever.

In this case, we would also have to add a label indicating:

WARNING! THIS EQUIPMENT FED BY MULTIPLE SOURCES

TOTAL RATING OF ALL OVERCURRENT DEVICES, EXCLUDING MAIN SUPPLY OVERCURRENT DEVICE, SHALL NOT EXCEED AMPACITY OF BUSBAR

18. Which of the following scenarios would most likely NOT comply with rapid shutdown of PV systems on buildings for utility-interactive systems?

- **a.** Microinverters on a hospital
- **b.** **Dc-to-dc converters that take two modules in series on and industrial rooftop**
- **c.** BIPV with no metal parts on a roof with no metal parts with a contactor disconnect inside the array
- **d.** A ground mount with an inverter mounted on a residential wall with a contactor disconnect between the ground mount and the wall.

We should look to **690.12(B)(2): inside the array boundary** to answer this question, which is what many people refer to as module level shutdown. 690.12(B)(2) became effective 1/1/2019.

Most of the time 690.12(B)(2)(2) will be the way module level shutdown is satisfied, which will typically be with module level power electronics (MLPE). MLPE includes microinverters and dc-to-dc converters, however if, a dc-to-dc converter is connected to two typical modules that are connected together in series, then it is really not "module" level. 690.12(B)(2)(2) limits voltage to 80V after shutdown is initiated. With most modules, your cold temperature corrected Voc is going to be over 40V, so two in series would take you over 80V. This is why the 80V number was put in place. The correct answer is the answer that would not work, which is **b.**, two modules in series on a dc-to-dc converter.

690.12(B)(2)(3) was put into place so that BIPV would be allowed in the NEC and, in this case, we cannot have exposed wiring methods, conductive metal parts, neither can we be within 8 feet from exposed grounded conductive parts or ground. 690.12(B)(2)(3) is why the answer **c.**, can be an acceptable installation.

Answer **d.**, can be acceptable because rapid shutdown is only required on buildings and if the wiring on the building can be shutdown with a contactor, then it is acceptable.

19. If a PV module has anodized aluminum frames and is installed with anodized aluminum rails, what is true about grounding or bonding of the modules?

- **a.** The grounding from the array to ground must have a continuous or irreversibly spliced grounding conductor to earth.
- **b.** If it is an ungrounded array, then you do not need to ground the array.
- **c.** **If you have a letter from the manufacturer saying you can use WEEB bonding washers and the bonding washers are UL listed, you CANNOT use them unless bonding washers are mentioned in the installation manual for the modules.**
- **d.** Copper equipment grounding conductors CANNOT be used in a system with aluminum PV module frames.

Equipment grounding conductors (EGCs) do not have to be continuous or irreversibly spliced. Grounding electrode conductors (GEC) do have to be continuous or irreversibly spliced for system grounding requirements. System grounding is when a current-carrying conductor is bonded to ground, which can only happen in a maximum of one place per system. With system grounding, there will be a white (or gray) grounded conductor or at least a white-marked grounded conductor.

Ungrounded PV arrays do have equipment grounding requirements if they have metal frames. The only time a module will not have equipment grounding requirements is when there are no exposed metal parts on the module. A module without a frame is called a PV laminate.

If the installation manual does not have instructions for using bonding washers, such as WEEBs in the section on grounding methods, then you CANNOT use them. When the module was tested by UL, then the Underwriters Laboratory will follow the instructions for testing according to the installation manual. If it is not in the manual, then you are not allowed to use bonding washers. Every day many people install modules with bonding washers that were not tested for use with bonding washers and they often get away with it. It is always up to the AHJ or inspector. If you argue with an inspector they can tell you to ground your array according to the module manufacturers' instructions. **NEC 110.3(B) says that we have to follow manufacturers' instructions**. Therefore, answer **c.** is the correct answer. **Always look closely at the wording of a question and do not get tripped up by negative NOTs and CANNOTs.**

Copper conductors are used in PV systems to bond metal frames of PV modules, although the bonding connection is made through an intermediate metal, such as a stainless steel piece of hardware or a tin-plated copper lug.

20. On a PV system, one of your workers said that he touched a white wire in a combiner box and got shocked. What is the most likely reason for this? The system was installed in 2007.

- **a. Ground fault**
- **b.** Positive to negative short
- **c.** Copper equipment grounding conductor touching aluminum on array
- **d.** The installer was hit by lightning

There is a label on a dc disconnect of a "retro" fuse-grounded type of functional grounded (formerly known as grounded) PV system that says that, **if a ground fault is indicated, normally grounded conductors can be energized with reference to ground.** This label is only required to be present if the system was installed before the 2017 NEC was adopted. If a dc wire was white, there should be a fuse-grounded inverter with a transformer if installed before the 2017 NEC was adopted. Also, almost every inverter installed in the USA in 2007 was with a fuse-grounded inverter. A non-isolated type of functional grounded inverter, formerly known as an ungrounded or transformerless inverter should not have a white wire. Additionally a fuse-grounded inverter installed after the adoption of the 2017 NEC should not have a white wire on the dc side, since we no longer consider these inverters solidly grounded. If the ground-fault circuit interrupting (GFCI) fuse in an inverter is blown, then there will not be a connection between the white wire and ground. Answer **a.** is correct.

With a short, the array would go to Isc and the voltage between positive and negative would be 0 in a dead short. In some cases, it could be argued that shorting an array could increase safety, since at Isc there is no voltage.

If the copper touched the aluminum on an array, the problem there would be that the aluminum could corrode, which is why we do not connect aluminum to copper due to galvanic corrosion of dissimilar metals.

The installer being hit by lightning could definitely cause a shock, but lightning is not that common, and the installer would probably not come to you complaining of a shock.

21. A house has a 200A panelboard and a 200A busbar with 220A of loads already in the panelboard. In which situation can you install a 40A solar backfed breaker on the load-side inside the panelboard?

- **a.** If the solar breaker is marked line and load.]
- **b.** If there is another 40A backfed breaker on the other end of the center-fed panelboard.
- **c.** **If the busbar is center fed and there are no other solar breakers installed.**
- **d.** Only if there are no quad or dual breakers doubling up on breaker spots.

According to 705.12(B)(2)(3)(B), otherwise known as the 120% rule, the sum of the main supply breaker feeding the busbar and 125% of the current of the utility-interactive inverter can equal 120% of the rating of the busbar, as long as the main and solar breakers are at opposite ends of the busbar. **With a center-fed main breaker, we can apply the 120% rule, which was not always the case.**

When applying the 120% rule, there are no requirements for looking at loads, whether they are continuous or non-continuous. In fact, adding solar to an overloaded busbar can actually help the busbar, by feeding the loads from the opposite end. It can be compared to shedding loads. There are no NEC rules against quad or dual breakers, but there are rumors that some AHJs do not allow backfeeding them. A quad breaker is sometimes used when there are no more breaker spaces left, you can then pull out a two-pole breaker and replace it with a quad, which is essentially two two-pole breakers that fits in one two-pole breaker space.

If a breaker is not marked line and load, it can be backfed.

22. At least how deep must PVC conduit be buried in a trench across a dirt yard with PV source circuits in the conduit?

- **a.** 12"
- **b.** **18"**
- **c.** 3 feet
- **d.** 6 feet

According to **NEC Table 300.5: minimum cover requirements**, typical non-metallic conduit must be buried at least 18". If it were buried under a building, concrete, etc., there would have been different requirements according to Table 300.5.

To look this up, you could go to **Article 352: rigid polyvinyl chloride conduit**, Type PVC, and look to 352.10(G): underground installations, where you will be directed to Table 300.5.

23. You can fit and sell 24 modules on west- and south-facing roofs. What would be the best combination for maximum production? Your single inverter has one MPPT.

- **a. 12 modules on the west and 12 on the south**
- **b.** 19 modules on the south and five on the west
- c. 13 modules on the south and 11 on the west
- **d.** 24 modules on the west and zero facing south

When putting source circuits together on an inverter with a single MPPT, all PV source circuits (strings) should have the same orientation.

With 12 on the south and 12 on the west, there can be two source circuits of 12 each, which is good. It would be better if there were 24 modules facing south, although the roof may not fit all 24 facing south.

If there were 19 facing south and five facing west, there are no combinations where there could be PV source circuits of the same length on both roofs. 19 is a prime number and there is no number except 1 that would go into 19 or 5. This would be a bad combination for an inverter with a single MPPT.

If there were 13 facing one direction and 11 the other, that would be another case of prime numbers and no possibilities for multiple source circuits of the same lengths facing different directions.

With all 24 modules facing west, it would be a good design; however, in the northern hemisphere it would be better to have half of the modules facing south. It could be argued that under some circumstances, such as a time of use utility rate schedule, that west facing can be as good as south; however, **in most circumstances the best answer would be 12 facing south and 12 facing west.**

Another benefit of 12 facing south and 12 facing west is that the peak production would be distributed over more hours and it would put less pressure on the inverter at noon and possibly prevent inverter clipping. Inverter clipping is when an inverter is maxed out and cannot make as much power as the array could potentially make. This would happen often if we put 6kW on a 4kW inverter and faced the array all in one direction.

24. Which is the best type of ladder to use when working with electrical equipment and for getting onto a roof?

- **a.** Aluminum ladder
- **b.** Painted wooden ladder with no metal parts
- **c.** **Fiberglass ladder with aluminum steps**
- **d.** A-frame fiberglass ladder with fiberglass steps

Aluminum ladders conduct electricity. **A painted ladder can hide cracks** and should not be used. A-frame ladders are not for climbing onto roofs. **Fiberglass ladders are used by electricians and typically have aluminum steps.**

25. Rigid non-metallic conduit is going 250 feet in a location where the low temperature is −12C and the high temperature is 38C. How many expansion joints must be used? Each expansion fitting can travel 4".

- **a.** 1
- **b.** 2
- **c.** 4
- **d.** **3**

In Article 352: rigid polyvinyl chloride conduit: Type PVC, in Section 352.44: expansion fittings, it says that expansion fittings are required when expansion is 1/4" or greater. Since the low temperature is −12C and the high temperature is 38C, then the difference in temperature is:

$$-12C - 38C = \mathbf{50C}$$

Table 352.44 shows temperature changes in inches per 100 feet as a product of degrees F or as degrees C in millimeters per meter. We can either convert metric length to imperial units or we can convert Celsius to Fahrenheit.

Since a 1° change in Celsius is equal to 1.8° change in Fahrenheit, then

$$1.8F \text{ per } C \times 50C = \mathbf{90F\ change\ in\ temperature}$$

In Table 352.44, a 90F change in temperature corresponds to 3.65" of expansion per 100 feet of PVC.

Since we have 250 feet of conduit, then:

250 feet × 3.65"/100ft = **9.13" of expansion**

Since each expansion joint can accommodate 4" of expansion:

9.13"/4" = **2.28 expansion joints**

Since rounding-down would not be enough, we must round-up, so **three expansion joints** are required.

26. In a positively grounded, solidly grounded direct water pumping PV system, what colors should the conductors be?

- **a. White positive, black negative, green ground**
- **b.** Red positive, black negative, bare ground
- **c.** Black positive, white negative, green ground
- **d.** Red negative, black positive, bare ground

Section 200.6: means of identifying grounded conductors tells us that the grounded conductor must always be white, gray, or have three continuous white or gray stripes or marked white.

Since this system is solidly positively grounded, the positive would be the answer with the white wire.

In previous versions of the NEC, a fuse-grounded PV system was required to have a white wire, but as of the 2017 NEC, much of grounding terminology has changed and only solidly grounded PV systems will now have a white grounded conductor.

27. In a residential 240V ac split-phase service, what is the relationship of the conductors to each other?

- **a.** The white wire is 120V from ground.
- **b. Line 1 and Line 2 are both 120V from ground.**
- **c.** Line 1 is at ground voltage.
- **d.** Line 2 is at ground voltage.

With residential 120/240V split-phase service, **Line 1 and Line 2 are both 120V from ground**, alternating in opposite directions from each other. The white "neutral" or "grounded conductor" is bonded to ground in one place. The white wire is 0V to ground, while L1 and L2 are both 120V to ground. Since L1 and L2 are alternating in opposite directions, they are measured at 240V when measuring L1 to L2 voltage.

In other countries, you will find L1 at ground voltage, but it will not be split phase.

If we were talking about 120/208V three-phase power, then the measurement to L1 or L2 to ground would be 120V, but when L1 and L2 (or L3) were measured, the resulting voltage would be 208V. This is because with three phase, the different lines are not alternating in opposite directions. They are considered to be 120° out of phase, rather than 180°. To calculate three-phase differences, we multiply by the square root of 3, which is about 1.73. Therefore, 120V × 1.73 = 208V. Also there are benefits with current when doing calculations: you benefit by being able to decrease your wire size by the magnitude of 1.73.

28. What is the best configuration for connecting lead-acid batteries, assuming that all of the combinations below add up to the same amount of energy in the battery bank and that the battery bank is 48V nominal voltage?

- **a. 24 batteries in series.**
- **b.** Two sets of 12 batteries in series in two different circuits paralleled together.
- **c.** Four sets of six batteries in series in four different circuits paralleled together on large busbars.
- **d.** Eight sets of four batteries in series paralleled together.

When configuring a lead-acid battery bank, it is best to have the **least amount of parallel connections** and, ideally, we would have a single string of batteries, so that there are no parallel connections.

The reason we want to avoid parallel connections is because if one string of batteries had less resistance, then that string would be used more than the other strings and would wear out sooner. Since batteries are essentially a chemical soup, it is very likely that this would happen. Also, if one conductor was longer or had a loose connection, then there would be less resistance on that string of batteries and the other string with less resistance would get used up sooner. Additionally, large battery banks with many parallel connections can have parasitic currents and use up energy even without loads.

29. What is the minimum size of an equipment grounding conductor in conduit for a PV source circuit of 12 modules in series with modules? There are no fuses or overcurrent protection devices on the source circuit. The following specifications of the modules are: Voc 44V, Vmp 35V, Isc 8.5A, Imp 7.8A.

- **a.** **14 AWG copper**
- **b.** 12 AWG copper
- **c.** 10 AWG copper
- **d.** 6 AWG copper

According to Section 690.45, when there is no overcurrent protection device is used, we should size an assumed overcurrent device based on 690.9(B), which is based on 156% of Isc. (In the 2014 NEC we were using 125% of Isc.)

$$8.5\text{A} \times 1.56 = \mathbf{13.3A}$$

According to Table 250.122, a **14 AWG copper** conductor would be sufficient for up to 15A.

30. A charge controller has a maximum input voltage of 150V and a minimum recommended operating voltage of 65V. The module you are using has the following specifications: Voc 44V, Vmp 35V, Isc 8.5A, Imp 7.8A, temperature coefficient for Voc = -0.34%/C, temperature coefficient for Vmp = -0.48%/C. The location has a high design cell temperature of 60C and a low design ambient temperature of -22C. What would be the possible source circuit configuration(s)?

- **a.** Two or three in series
- **b.** Only three in series
- **c.** Three or four in series
- **d.** **Only two in series**

First, we will determine the longest PV source circuit using the low temperature, the Voc, the temperature coefficient of Voc, and the high voltage input of the charge controller.

The difference between the cold temperature and STC, which is how the module was tested in the lab, is:

$$-22\text{C} - 25\text{C} = \mathbf{-47C}\ \Delta T$$

We multiply the change in temperature by the temperature coefficient of Voc:

-47C × -0.34%/C = **16.0% increase in Voc**

To get 16% more voltage than 44V, we can multiply:

44V × 1.16 = **51.0V when cold**

To determine the maximum number of modules in series we divide the voltage of the cold module into the maximum input voltage of the charge controller:

150V/51V = 2.94 modules in series

Since we cannot have a fraction of a module, we will have to **round-down to two modules in series** so we do not go over voltage and damage the charge controller.

To determine the minimum number of modules in series, we use the Vmp, the hot temperature, the temperature coefficient for Vmp and the low-end voltage of the charge controller.

The difference of the hot temperature from the temperature that the modules were originally tested at is:

60C - 25C = **35C** ΔT

We will multiply the difference in temperature by the temperature coefficient of Vmp:

35C × -0.48%/C = **−16.8% decrease in Vmp**

We can subtract 16.8% from 100% to determine the amount of voltage that we keep after our losses:

100% - 16.8% = **83.2% of Vmp remaining at hot temperatures**

We can turn the 83.2% into a fraction to derate the Vmp:

0.832 × 35 Vmp = **29.1 Vmp when hot**

Now to find our smallest string size, we will divide our low voltage of 29.1 Vmp into our low-end charge controller voltage of 65V

65 V / 29.1 Vmp = **2.23 in series**

Once again, we cannot use a fraction of a module, but this time we will round-up, since rounding-down would give us a voltage that was too low. In this case, the least amount of modules in series would be two modules.

With this charge controller, module and temperature combination, **we will only be able to have PV source circuits of two in series**. Having any less would not be enough on a hot day and having any more would be too many on a cold day.

Strategy

Chapter 14

Short-term memory tips:

1. Read the contents and index of the NEC the night before the exam.
2. Review practice questions the days before the exam.
3. Read Sections 690, 705, 706, and other relevant sections of the NEC two days before the exam.
4. Practice calculator and calculations days before the exam.
5. If your time is very limited, study practice exam questions in this book.
6. Find where the exam will be held and where you will park or how you will get there before taking the exam. It would also be a good idea to study there if possible.
7. State dependent memory: In psychology classes, you may have learned about state dependent memory, which says you will do better if you study under the same conditions that you take the exam. Study at the same time of day and under the same conditions that you will be taking the exam.
8. Practice with an NEC Codebook that you will bring to the exam if you are allowed.
9. Do not change your caffeine habits for the exam. Having withdrawal symptoms from not having drunk enough or being too wired can cause focusing problems. Do not worry about falling asleep during the exam; it is too exciting to do that.
10. Look for clues within other exam questions that will help you along the way and stay sharp, so you can go back quickly.
11. Take notes on the exam and have a system that will help you when you go back. Mark questions that you were sure you knew differently than questions that you thought you might know differently than the questions that you guess on.
12. Pay attention to the time. Since it is a 4 hour exam, you should be at least halfway finished after 2 hours. Pace yourself and try to leave 15 minutes at the end to review everything.

13. Never leave the exam early if you are serious about passing. Most people leave early and many people who failed by one question could have passed if they were not so excited about leaving early.
14. Get as much sleep as possible.
15. Eat a good breakfast.

Other recommended reading in your "spare time":

- *National Electrical Code Codebook.*
- *Photovoltaic Systems and the National Electrical Code*, by Sean White and Bill Brooks.
- *Solar Professional* magazine archives.
- *NABCEP PV Installation Professional Resource Guide*, by Bill Brooks and Jim Dunlop.
- *Photovoltaic Systems*, by Jim Dunlop.
- *Photovoltaic Power Systems for Inspectors and Plan Reviewers*, by John Wiles.
- *Understanding Requirements for Solar Photovoltaic Systems*, by Mike Holt.

Figure 14.1 Get to know the NEC

For updates and corrections go to www.solarsean.com

Index

Page numbers in italics refer to figures. Page numbers in bold refer to tables.

Taylor & Francis eBooks

www.taylorfrancis.com

A single destination for eBooks from Taylor & Francis with increased functionality and an improved user experience to meet the needs of our customers.

90,000+ eBooks of award-winning academic content in Humanities, Social Science, Science, Technology, Engineering, and Medical written by a global network of editors and authors.

TAYLOR & FRANCIS EBOOKS OFFERS:

- A streamlined experience for our library customers
- A single point of discovery for all of our eBook content
- Improved search and discovery of content at both book and chapter level

REQUEST A FREE TRIAL

support@taylorfrancis.com